Deep in the Heart of Your Brain
notes from an exhibition

Jacqueline Donachie

ISBN 978 1 908638 27 4

British Library Cataloguing in Publication Data
A catalogue record for this book is available from the British Library.

Designed and typeset by Neil McGuire / After the News
Edited by Susan Pacitti
Printed in Scotland by Allander Print Ltd

Cover: GalerieArt Satin 300gsm FSC Mix Credit Cert. No. TT-COC-002590
Text pages to pp.57 GalerieArt Satin 150gsm FSC Mix Credit Cert. No. TT-COC-002590 / Text pp 57 – 197 Pop'Set Virgin Grey 120gsm FSC-TT-COC-002590

FSC www.fsc.org MIX Paper from responsible sources FSC® C008886

Publication of this book was made possible with grant funding from The Friends of Glasgow Museums and The Marigold Foundation.

Author acknowledgements:
Jacqueline Donachie would like to thank Professor Darren Monckton, Professor Chris Dorsett, Roderick Buchanan, Susan Craig, Kate V Robertson, Don Mackenzie, Patricia Fleming and all of the women who participated.

Contents

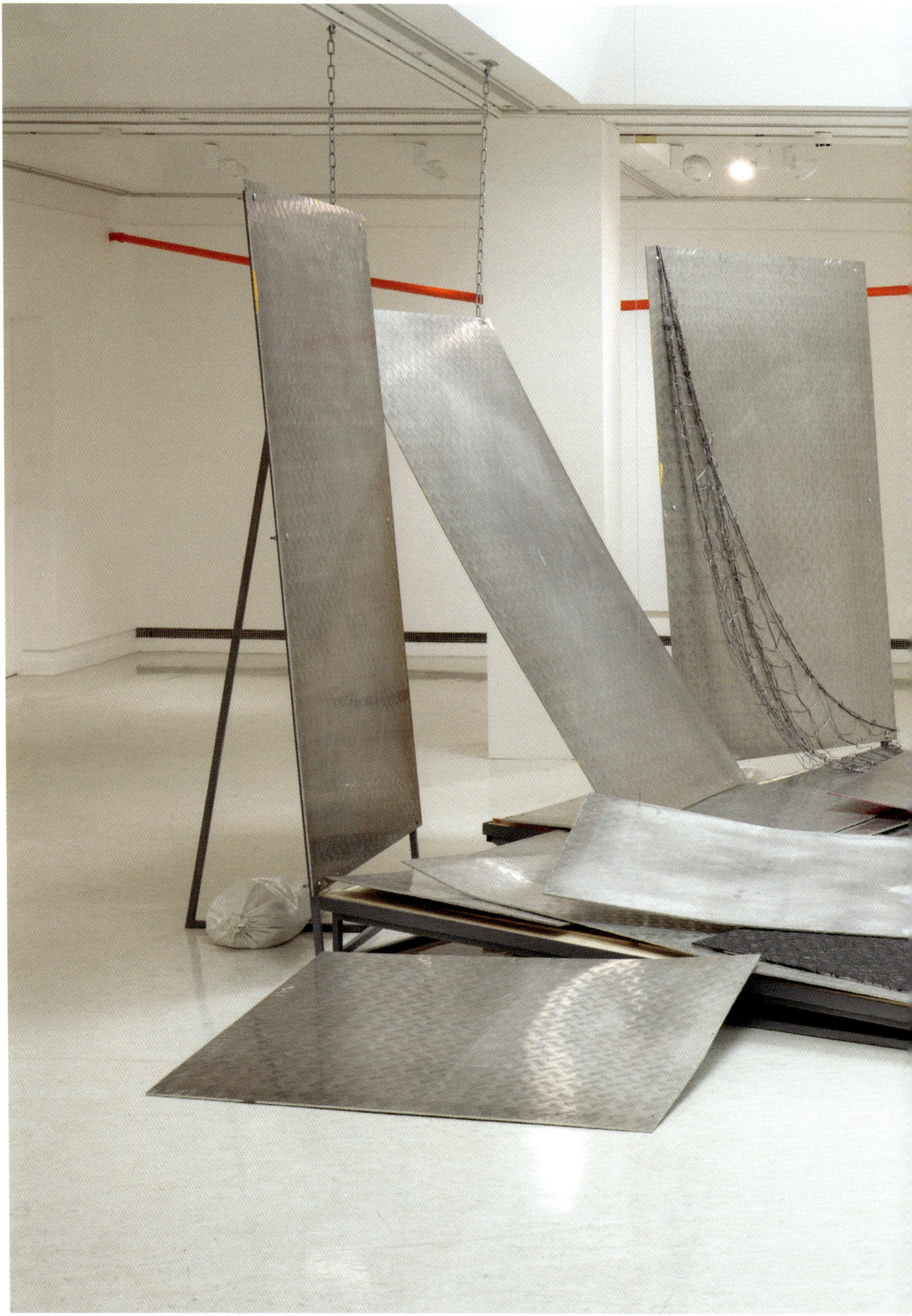

Installation view of *Headphone Music, Boats and Trains*, 2016, and *Nice Style*, 2016
(Glasgow Museums, S.482)

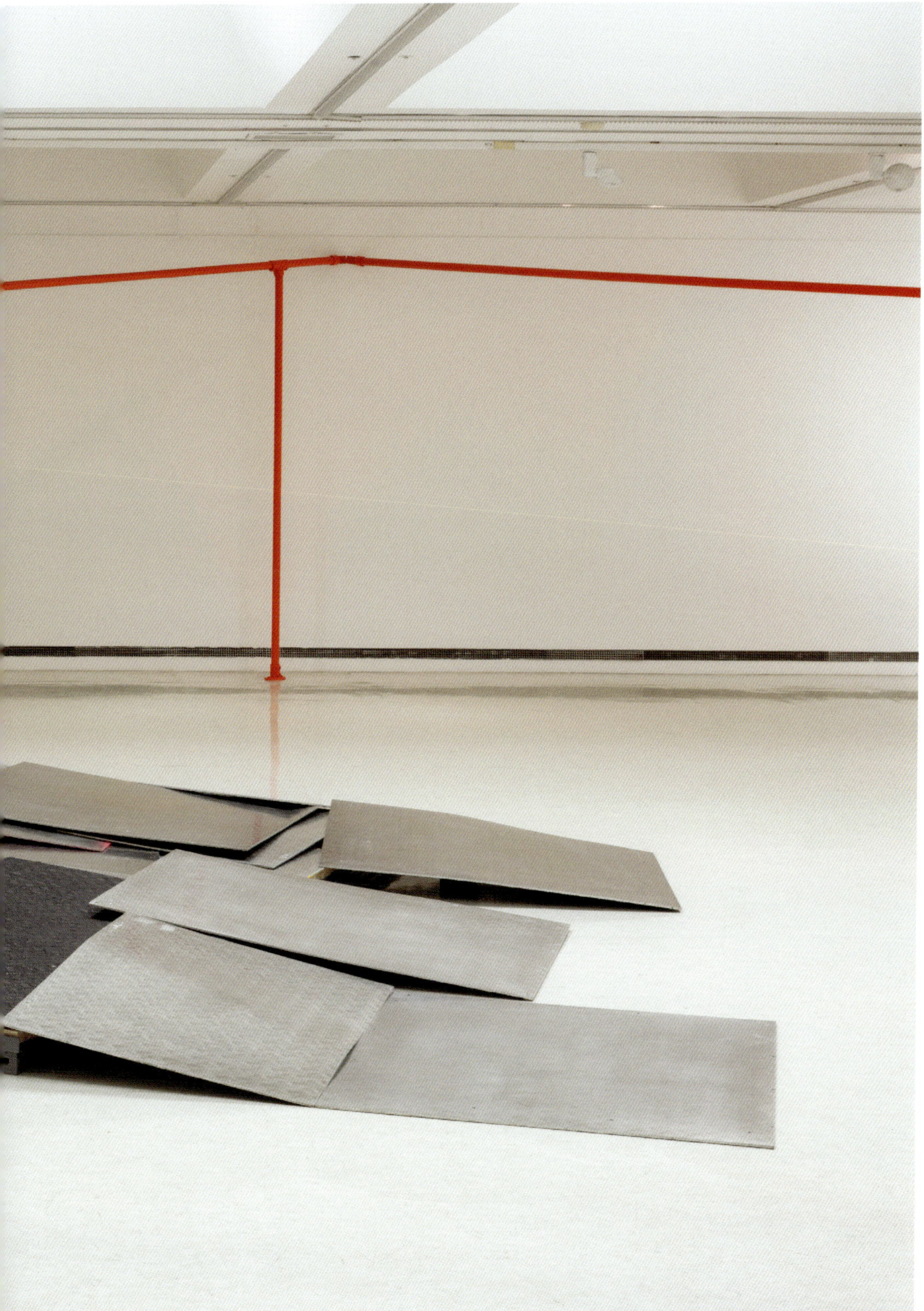

Foreword

The UK Myotonic Dystrophy Patient Registry has been thrilled to be part of the production of *Hazel* and Jacqueline Donachie's ongoing work. This is the first time the Registry has been used to recruit for research of any sort. In one sense it didn't matter that this was an art project rather than a clinical trial, as it provided an opportunity to show the purpose and function of the Registry. It instantly broadened the scope of the Registry, demonstrating its utility beyond the obvious and showing it as a tool which can facilitate research of any kind.

The John Walton Muscular Dystrophy Research Centre (JWMDRC) has worked with artists on a number of other projects. The Centre is a truly interdisciplinary unit, working in rare, often life-limiting, diseases, and is keen to engage in innovative strategies to increase knowledge and awareness surrounding its work. We have had the pleasure of seeing *Hazel* on a number of occasions, in front of a large audience at IDMC [the International Myotonic Dystrophy Consortium] and in the smaller setting of the clinical team at the JWMDRC. On both occasions we noticed the attention everyone paid to the film, no one checking emails or making notes, just watching. As professionals working with people affected by myotonic dystrophy, it was the first time we had had the condition presented to us in such a way. So often things are presented for academic or promotional reasons, but *Hazel* provided a very real, unfiltered view into the lives of these sisters. In addition, the full interviews gave a real voice to these often neglected patients.

On reflection, working with Jackie has made us consider how the medium of film and art in general can be used to make scientific research more accessible, but also to provide an insight into others' lives in a non-clinical and thoughtful way. Furthermore, it has allowed us to consider the wider scope and reach the Registry could have by collaborating with people from other fields.

Libby Wood, *UK Myotonic Dystrophy Patient Registry Co-ordinator*
Hanns Lochmüller, *Chair of Experimental Myology, Principal Investigator of the UK Myotonic Dystrophy Patient Registry*

Relationships: Public and Private

In 2011 I began a conversation with Jacqueline Donachie about working on something around health and art practice for the Gallery of Modern Art (GoMA). At that stage it wasn't a project or an exhibition but I was excited about working with her to develop a commission for the Gallery. Since 2002 Jackie has been collaborating with a range of scientific and medical professionals to produce new ideas and artworks exploring inherited genetic disabilities, wellbeing and care. Her work often stems from her family situation and a curiosity to know more about myotonic dystrophy (DM), the genetic condition that entered their lives. This growing body of work has gained a strong reputation in both art and medicine and compels us to see disability, family and care in new ways.

Jackie had already made a significant body of work on DM, including *Tomorrow Belongs to Me* (2006) with geneticist Darren Monckton, a film of interviews with key scientists involved in groundbreaking research to prove the phenomena of anticipation in certain genetic conditions (that is, where it is passed down through generations in increasingly severe or more prevalent forms). She was animated about an idea to make a further film on DM with sisters, mirroring her own family situation, but not sure how she would make contact with sisters, or find the funding to make the film. It took another four years before our discussions developed to the point of planning an exhibition, but in the meantime her research had progressed and she was in contact with sisters through the UK Myotonic Dystrophy Patient Registry at The John Walton Muscular Dystrophy Research Centre, Newcastle, as part of her PhD study.

This proposal to work with sisters really appealed to me. I don't have a sister – I have female cousins, friends, a mother and a daughter, but not a sister. This sibling relationship has fascinated me over the years at family gatherings, seeing cousins with their sisters; observing friends with their sisters, and now through this work with Jackie and her sister. Even though I am surrounded by women in my family, I have a quiet envy of a family bond I don't have, a fantasy about the sister relationship that is missing. In contrast, Jackie grew up with a sister but has seen their relationship change significantly through the circumstances of their genetic make-up.

It was a pleasure to see our initial discussions and Jackie's proposal evolve into a compelling body of work which was on one level about sisters and that relationship, but also spoke to a much broader conversation about care, relationships and the bewildering medical environments that we may unexpectedly find ourselves encountering in our own lives. These works were shown in Jackie's acclaimed solo show for GoMA, *Deep in the Heart of Your Brain* (Gallery 4, 20 May–13 November 2016). Central to this exhibition of newly commissioned sculptures, video and drawings was the film *Hazel,* made from the interviews with sisters contacted through the Registry. This is a powerful work, which engendered many conversations in the exhibition about siblings, ageing, loss and care. Full transcripts of the interviews were also available in the exhibition resource space, and we found that these were appreciated by visitors and by academics who wanted to hear more from all of the sisters. This curiosity of audiences to know more was one of the motivations for pursuing the idea of this publication.

Alongside the show we ran a public programme of talks, events and a symposium, examining the exhibition themes in a public forum. It was a privilege to work with Jackie, her collaborators in the science field, and partners, on a wider public programme raising the visibility of myotonic dystrophy, care and disability in a civic space in the heart of the city. Through her films, sculptures and drawings, Jackie has placed the lived experience and research of an inherited genetic condition at the heart of her work, whilst at the same time ensuring that same work remained distanced from the sentimental and the emotional. This mode of operating, and a reputation in both art and medical fields, is a space she has negotiated over many years.

I was delighted when we were successful in gaining funding from the National Fund for Acquisitions to acquire a selection of Jackie's work from *Deep in the Heart of Your Brain*. These are now in Glasgow Museums' collection, archiving a rich body of works about lived experiences, data and the aesthetic in dialogue with the city's fine art, scientific and social history collections. Her artwork inhabits a complex intersection of professional fields (art and science), but also has potential beyond the exhibition to engage wider audiences to think about disability, care and loss and to influence change for those lives affected. This publication intends to do the same by bringing together art, fiction and the lived experience.

It is fitting then, that this publication has emerged from that working relationship between myself, Glasgow Museums and Jackie as an excellent resource for scientists, academics, families affected by myotonic dystrophy, artists and museums. Alongside the full transcripts of the interviews with the sisters, we have commissioned a new work of fiction from author Nicola White. Nicola's only brief from us was a folder containing the original *Hazel* interviews and a request to think 'about a girl' – the resulting work has exceeded our expectations. We are also thrilled to have an essay from Moira Jeffrey placing Jackie's work in the context of others and looking at life stories, epiphany and sisters.

Glasgow Museums would like to thank The Marigold Foundation and the Friends of Glasgow Museums for their generous support of this publication, Jacqueline Donachie, Moira Jeffrey, Nicola White, and Libby Wood and Professor Hanns Lochmüller for reflections on the opportunities presented by a clinical registry working with an artist.

Above all I would like to thank the sisters Louise, Rachel, Jennie, Pam, Michelle, Stephanie, Lorraine, Sally, Lucie, Claire and Laura who gave their time and energy to the interviews and their permission to reproduce them in this publication. I hope that through this opportunity to read their lived experiences we can understand more about the resilience that disability demands, the complications of families, health and care, and what we can do in our respective fields to illuminate this and affect change.

Katie Bruce
Gallery of Modern Art, Glasgow

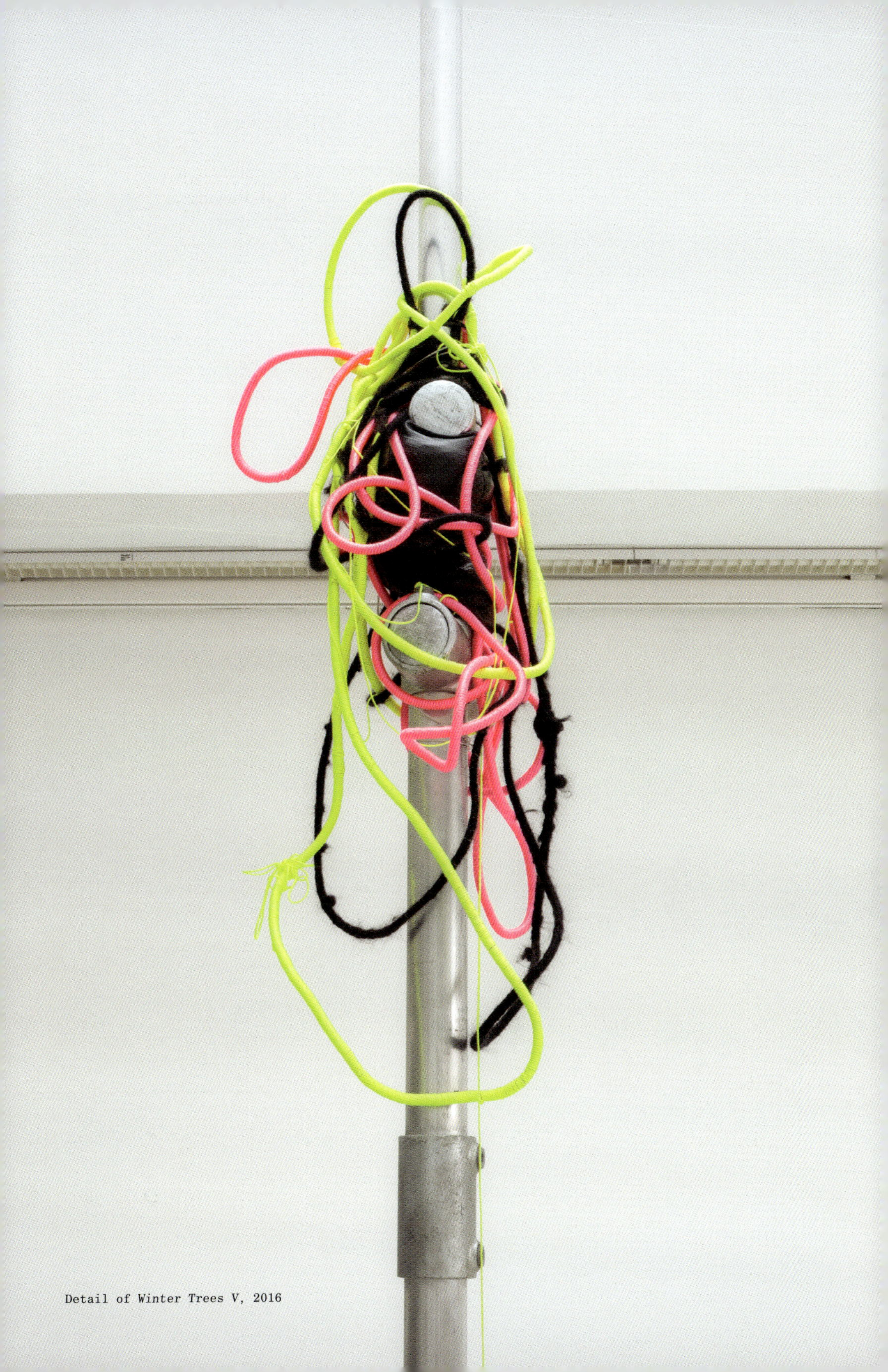

Detail of Winter Trees V, 2016

Hazel

Resemblance is a complicated thing. My entire first year at The Glasgow School of Art I was asked if I had a sister in second year. I was used to having a very-alike-could-be-twin-sister, but she wasn't at art school, didn't even live in the same city. Who was this other woman? When we eventually met, awkwardly, at the bar in the student union (of course) we both stood off; what was in me that they saw in her? Freckles, a long nose, blue eyes… In her friendly but quizzical look I saw her think the same. More than 20 years later we still occasionally meet around town and I wonder if others think we look alike even now, or has age affected our similarity?

I wonder too if I still look like my sister? This is an ongoing puzzle that has coloured my many years of research and collaboration with scientists. When do you stop looking like your family and start looking like the symptoms described in a textbook? When is familial resemblance overtaken by another layer of inheritance that, like long legs or short tempers, also comes from your parents? An inheritance that can cause all sorts of assumptions about health, ability and competence to be made both by 'the outside world' and within family dynamics.

The interviews for the artwork *Hazel* published here recognize the importance of considering the private, domestic experience of an inherited genetic condition by asking women to speak frankly about their own lives; I selected a very specific group for interview – sisters between the ages of 30 and 70, where one (or two) has inherited the myotonic dystrophy gene and one has not. This was to mirror my own position as an unaffected sister, and also to address the visibility of progressive female ageing, both in general and within the specific world of genetic research and care.[1] I have used these women, sisters, to portray a broad range of female experience connected by a single genetic factor, to illuminate the loss that affects each of us differently. There are certain shared aspects that directly connect to appearance – whether that be facial deterioration or the consideration of when, if at all, an affected sister becomes more recognized for her disability than her familial characteristics – but there is also much that still connects to individual life experience.

The women in the *Hazel* research are not poster girls for disability, nor even for their own genetic condition. They are quietly spoken women living with slow physical deterioration that's not just marked by wheelchairs or walking sticks, but by hair loss, collapsed facial muscles and low-income employment. But they are also daughters, friends, sometimes wives and sometimes mothers. They wear makeup, have tattoos, take holidays, cook dinner. Some are employed, some are not. Even when discussing highly emotional situations, such as infertility or the loss of a child, there is calm and there is resilience. Room for *Hazel*, and this research, to extend beyond a simple retelling of physical decline or collection of medical data into an opportunity to consider everyday activity, family relationships. To think of similarity, and what makes us look and act the same, and what divides and separates us beyond symptomatic observation.

In the film, two faces appear side by side. When a woman is seen alongside someone of a similar physical and genetic background who is also beginning to deal with the ageing process but does not have myotonic dystrophy, in a society filled with mediated images of what a woman in her 40s, 50s and 60s should look like, is the experience portrayed here useful to illuminate a loss that is difficult to define? Medicine references the moment a wheelchair arrives in the house (whether used or not) as a milestone in the condition, and the biographical disruption that causes young women to use walking sticks before their mothers, or grandparents to raise a toddler when their daughter is unable to. It doesn't always consider the lives lived beyond the clinic, and the personalities of those involved.

Collaborative process and the coalition of expert knowledge are essential tools for advancing research across many fields. Communication and observation are a cornerstone of the art-making process I was trained in, a process that at times overlaps in approach with other disciplines from genetics to social sciences. A refrain of anthropologist James Clifford's is his encouragement of anthropology's scientists to reflect on their relationship to the people/cultures they are studying. For example, he asks: 'What if someone studied the culture of computer hackers [...] and in the process never "interfaced" in the flesh with a single hacker? Would the months, even years, spent on the Net be fieldwork?'[2] Regardless of the answer to the question posed, it is hardly a stretch to see how this reflection on observer/participant methods can be transposed to the context of scientific/medical research and artistic practice: must researchers engage with the subjects of their inquiry? How do they do that? What are the advantages/limits of detached observation? What are the advantage/limits of personalized approaches? What insights would a collaborative approach that enables experts in different fields to come together somewhere between these two approaches provide?

These are points that the myotonic dystrophy community has considered through the structure of conferences and meetings in the past decade that have raised and acknowledged the expertise that families hold in the field. Patients are experts too, and art can be a powerful method of articulating lived experience. Everyone has a story to tell, to share, and it's on that basis that this book is published.

One point raised in the interviews is '*What if my sister didn't have it?*', something that, alongside a consideration of familial resemblance – of when siblings might start to look like their condition and not their sister – hangs like a cloud. Two women mulling over a what-if scenario that is difficult to voice. Is it OK to wish your family didn't have this? What would life be like with ordinary genes? Interestingly, all the women here were diagnosed as adults; in childhood no genetic condition was identified. But there were often differences: health problems, tiredness, an undefined lack of something; and likewise education and later employment. These are contemplations that have no answer in medicine or clinical fact and are not confined to myotonic dystrophy. Perhaps everyone with a sister wonders what it would be like to have a *different* sister, and those with no sisters wonder what it would be like to have one at all.

> *I don't think it's possible to answer the question if I feel I've lost something because I don't know what it would be that I've lost.*
>
> *… if you feel you've lost the sister that you would have had, if you had had her, you've lost the sister that you do have.*[3]

Jacqueline Donachie
July 2017

1. There is no scientific evidence of a difference in the manifestation of the condition in men and in women other than in terms of fertility and reproduction, and working with sisters only, as opposed to male siblings, stems from my personal situation.

2. James Clifford, *Routes: Travel and Translation in the Late Twentieth Century*, p.61. Cambridge, MA: Harvard University Press, 1997.

3. Interview with Lucie, 15 March 2015 (p. 165)

Deep in The Heart of Your Brain

Gallery of Modern Art
Glasgow, 2016

Hazel (3-screen installation, 2016)

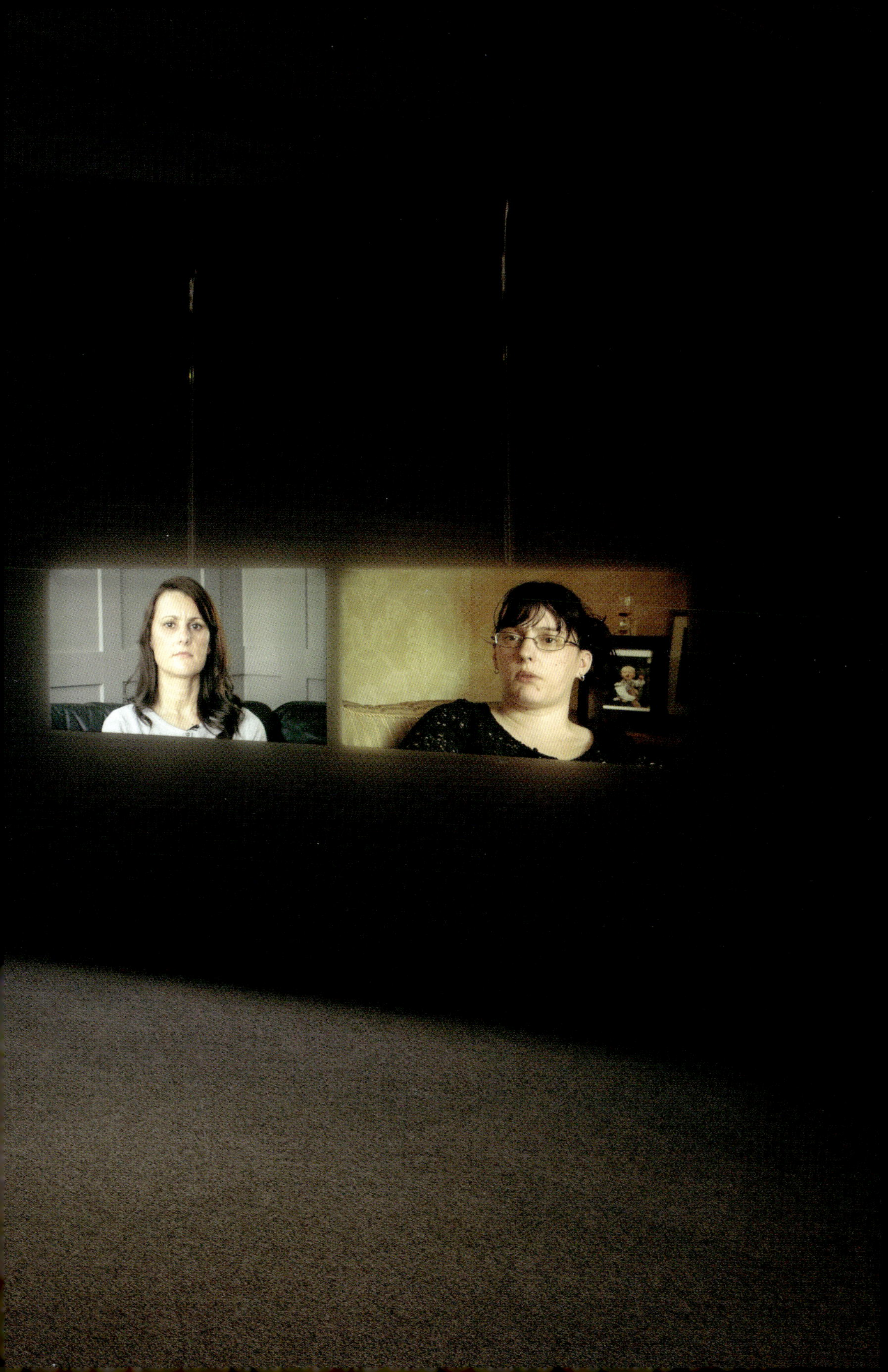

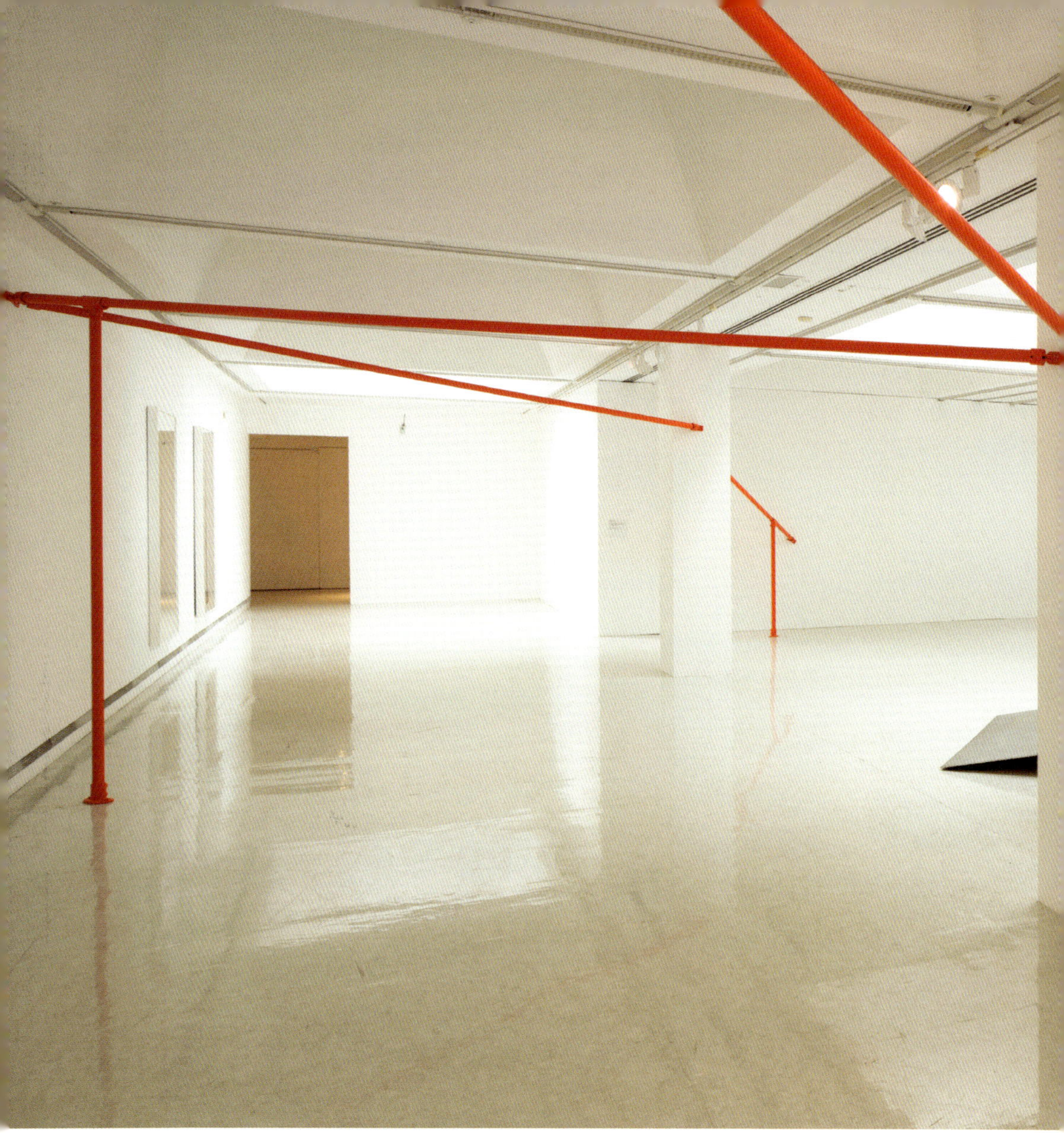

Installation view of *Headphone Music, Boats and Trains*, 2016, and *Nice Style*, 2016 (Glasgow Museums, S.482)

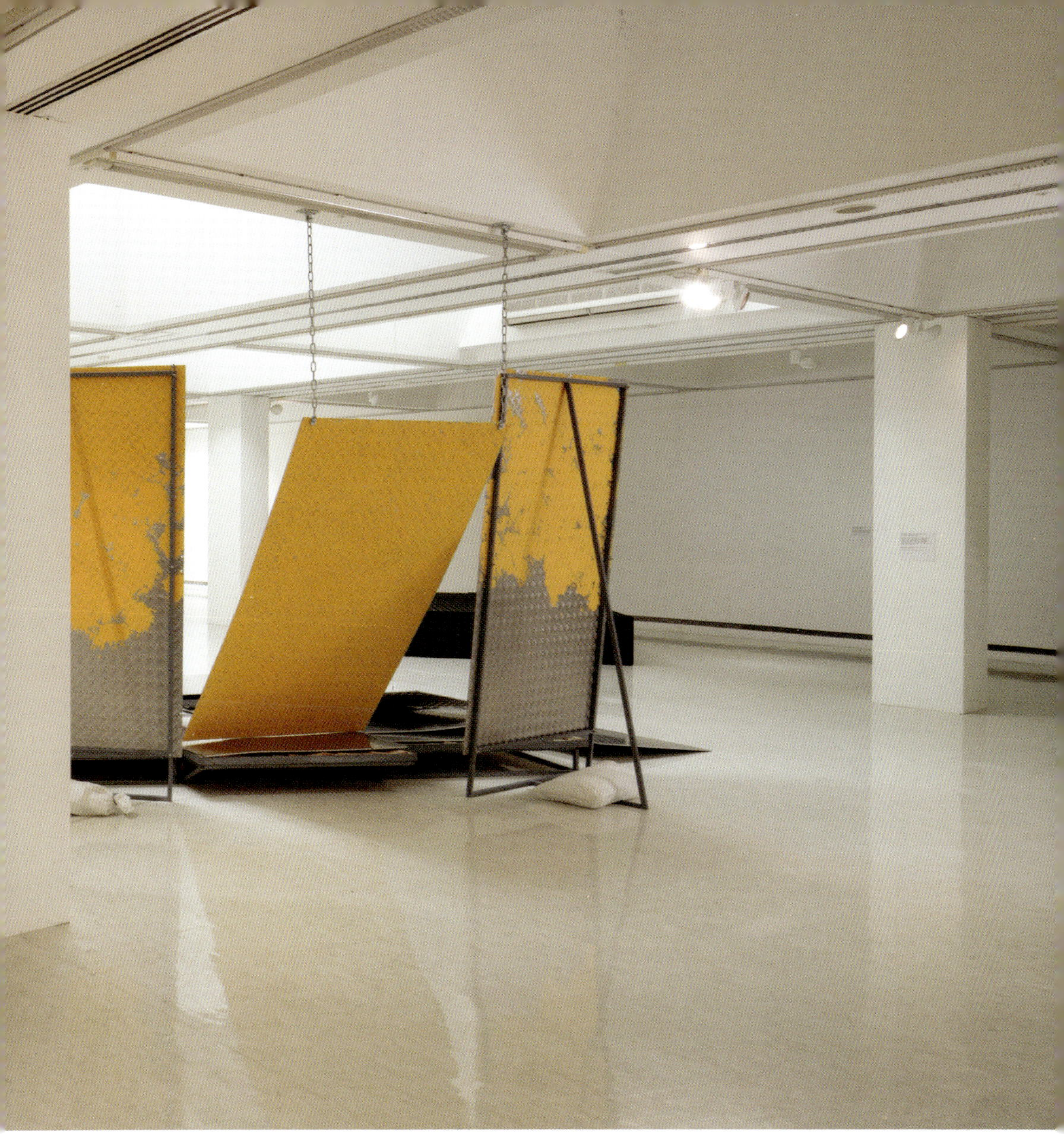

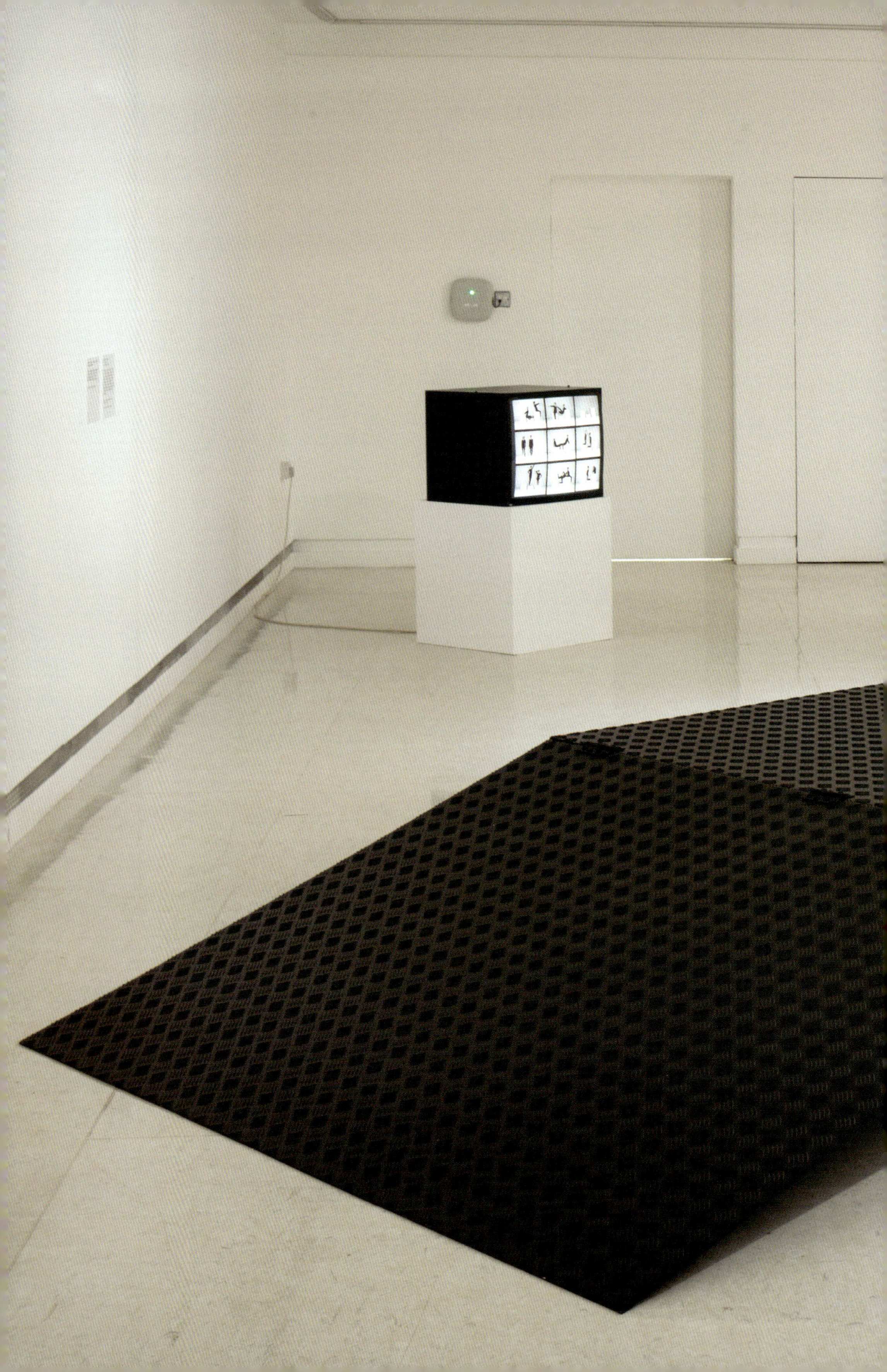

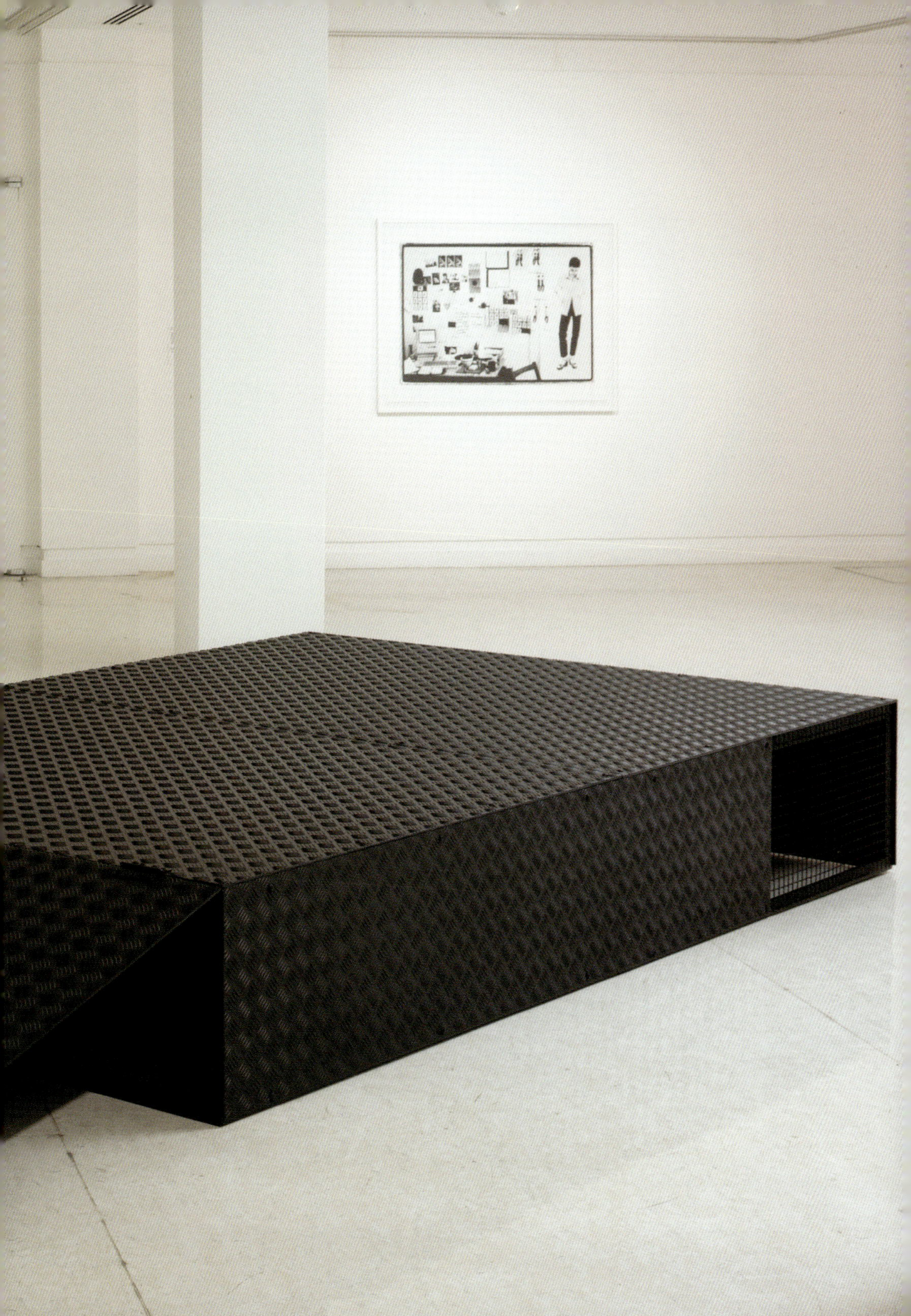

Installation view of *Deep in the Heart of Your Brain is a Lever*, 2014 (Glasgow Museums, S.483), *Pose Work for Sisters*, 2016 (Glasgow Museums, AV.2017.2), and *Studio 1995*, 2016 (Glasgow Museums, PR.2017.3)

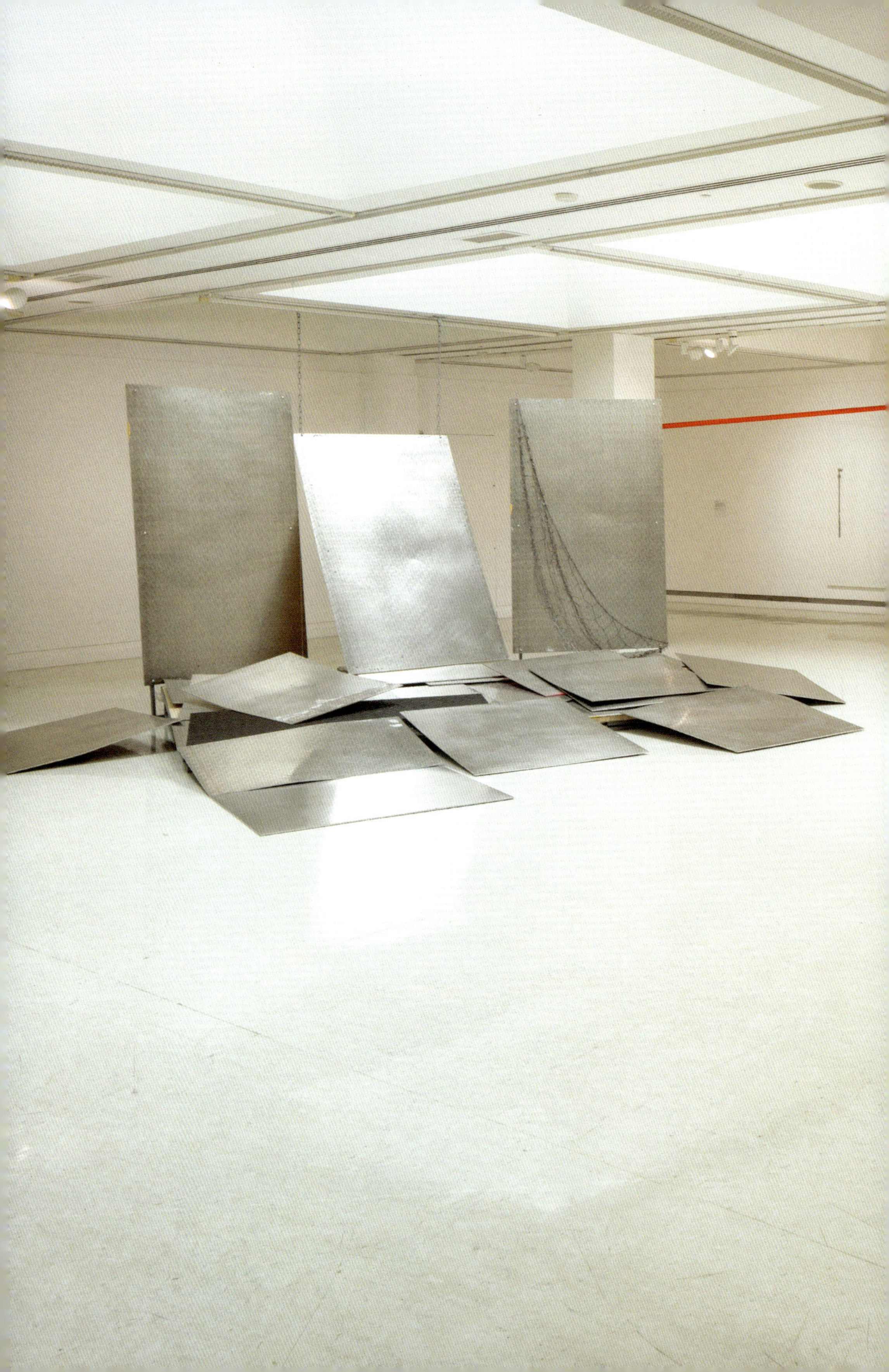

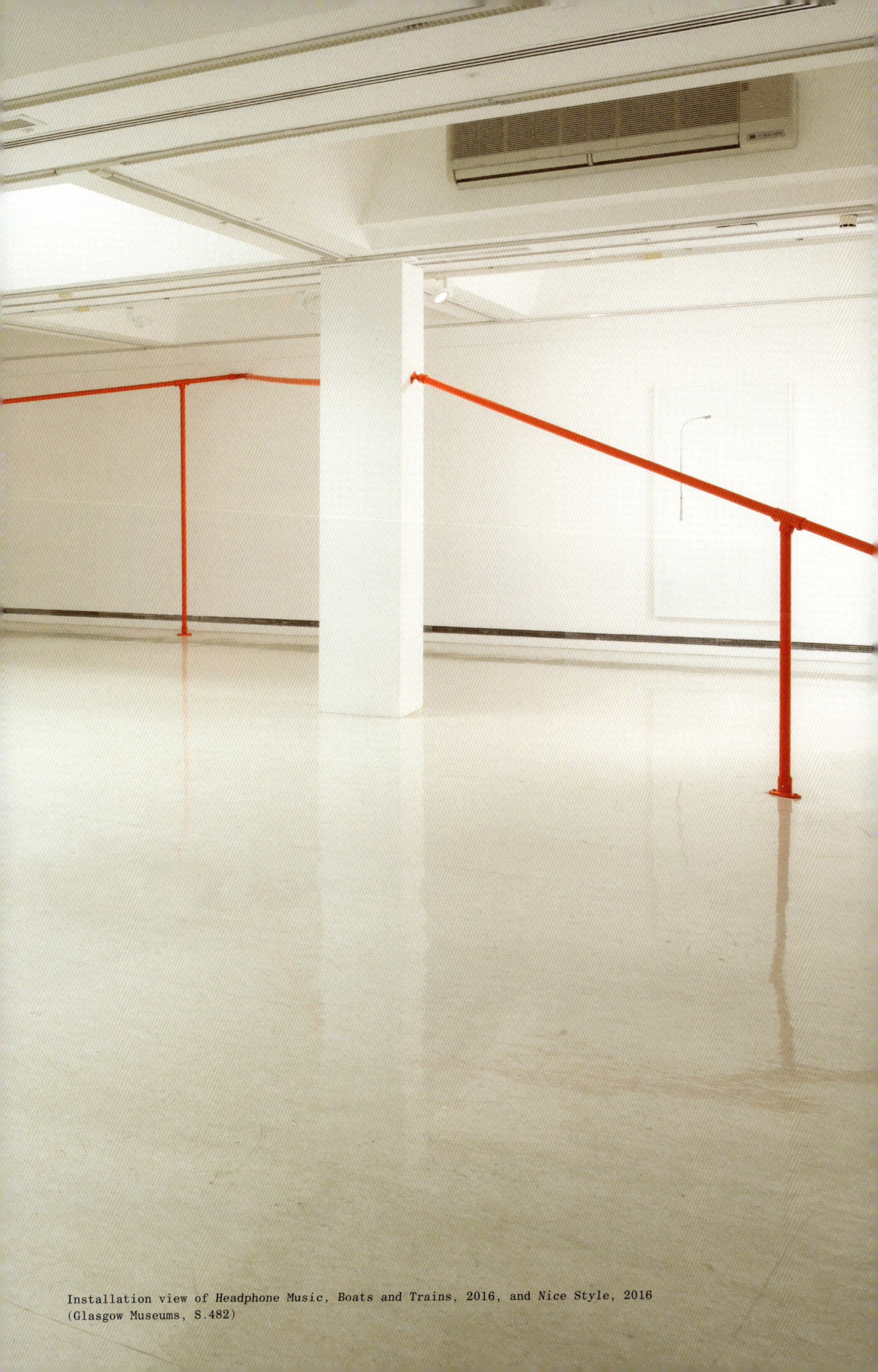

Installation view of *Headphone Music, Boats and Trains*, 2016, and *Nice Style*, 2016
(Glasgow Museums, S.482)

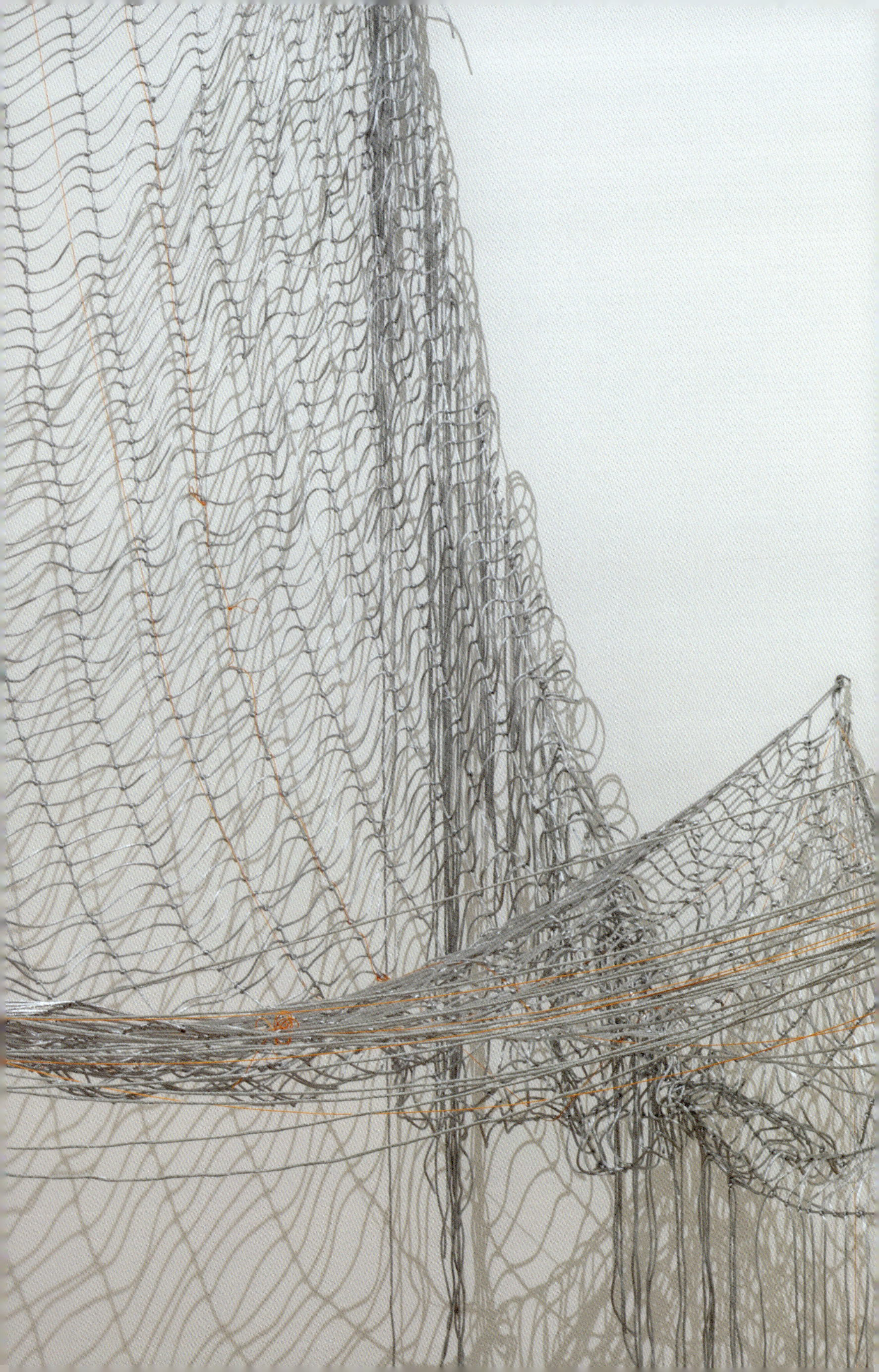

Detail of *Winter Trees IV*, 2016 (Glasgow Museums, S.484)

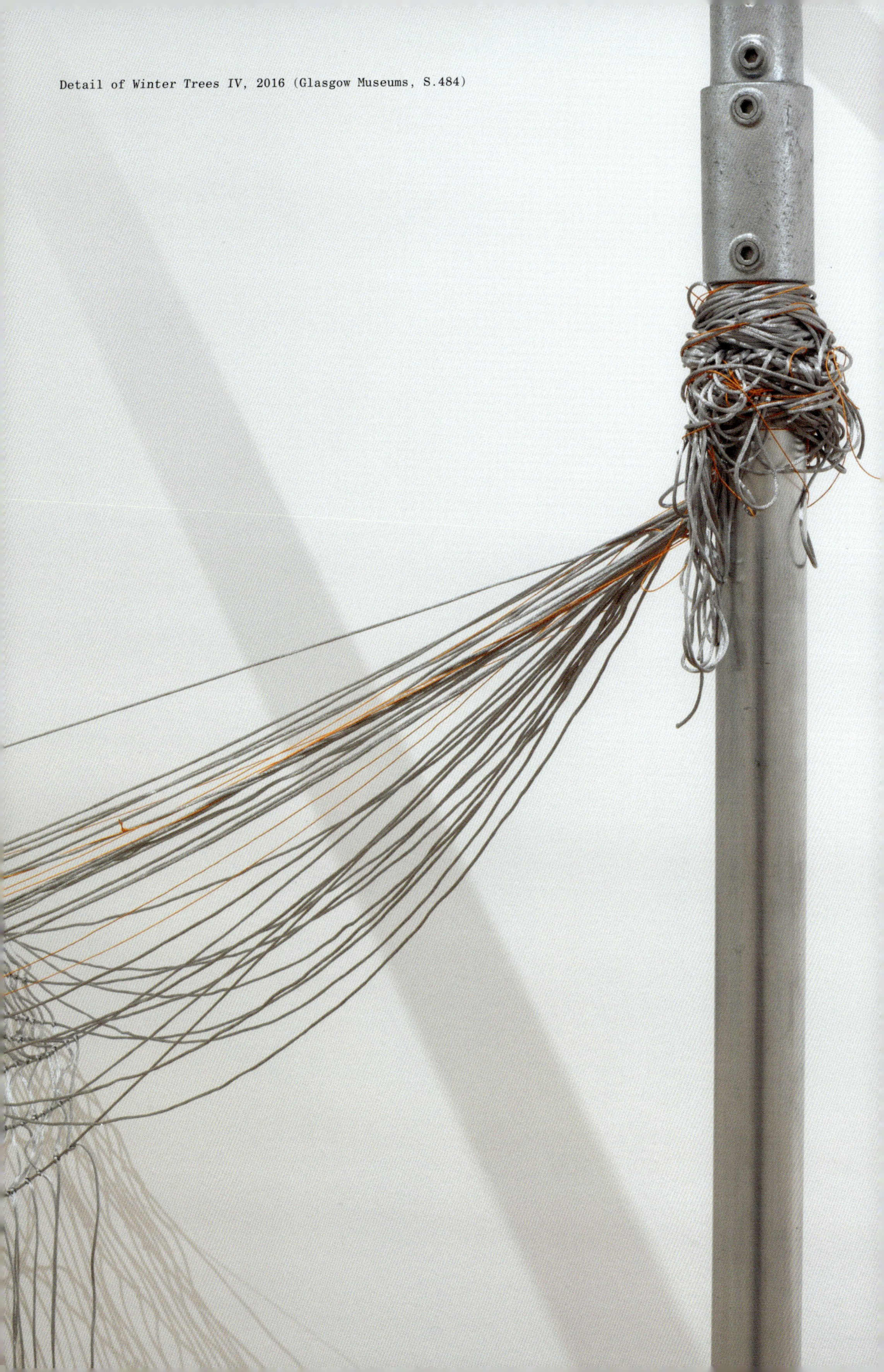

Detail of Winter Trees IV, 2016 (Glasgow Museums, S.484)

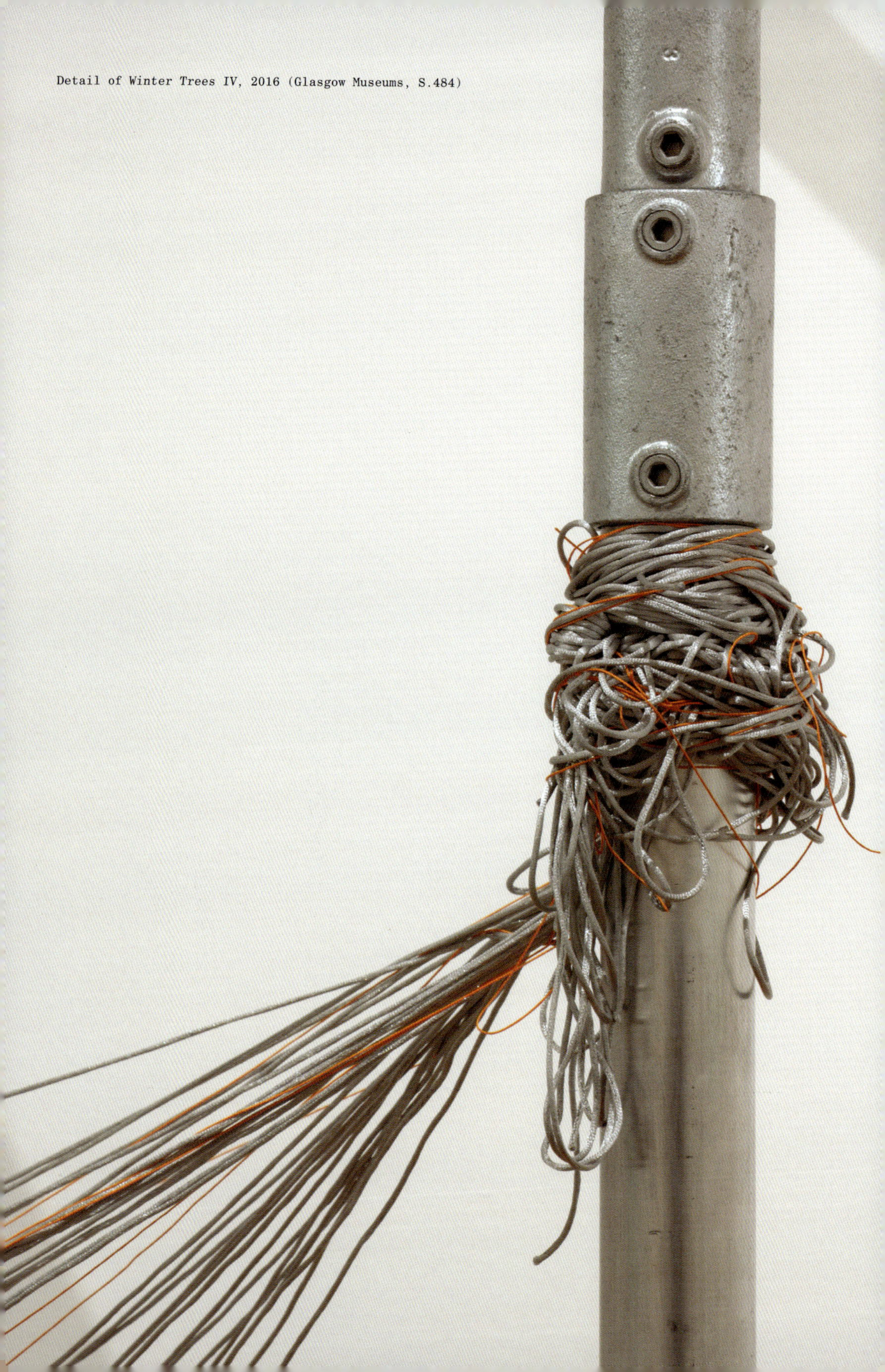

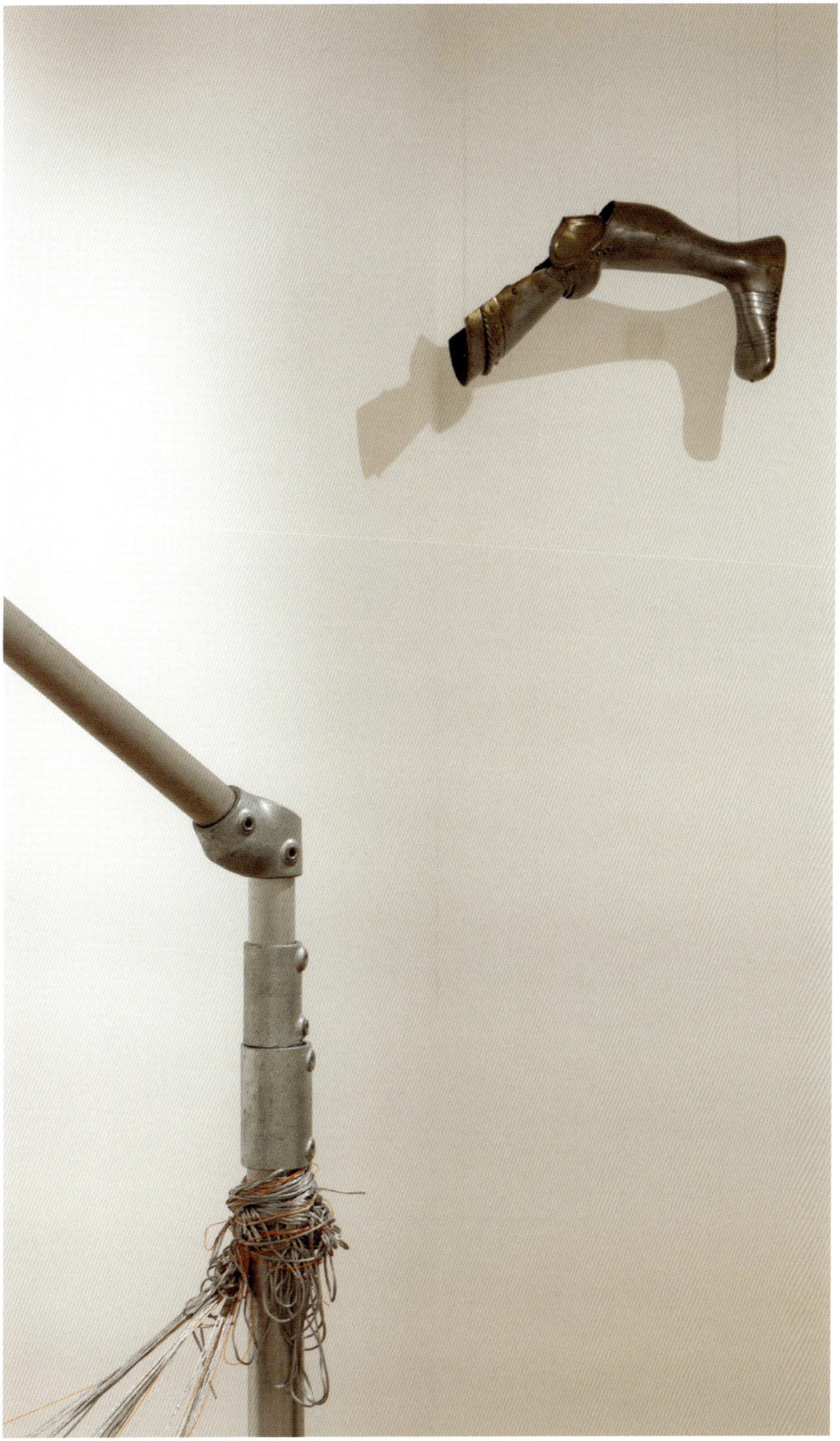

Detail of *Winter Trees IV*, 2016 (Glasgow Museums, S.484) and
Leg defence, 1500s (Glasgow Museums, RL Scott Collection, E.1939.65.u [11])

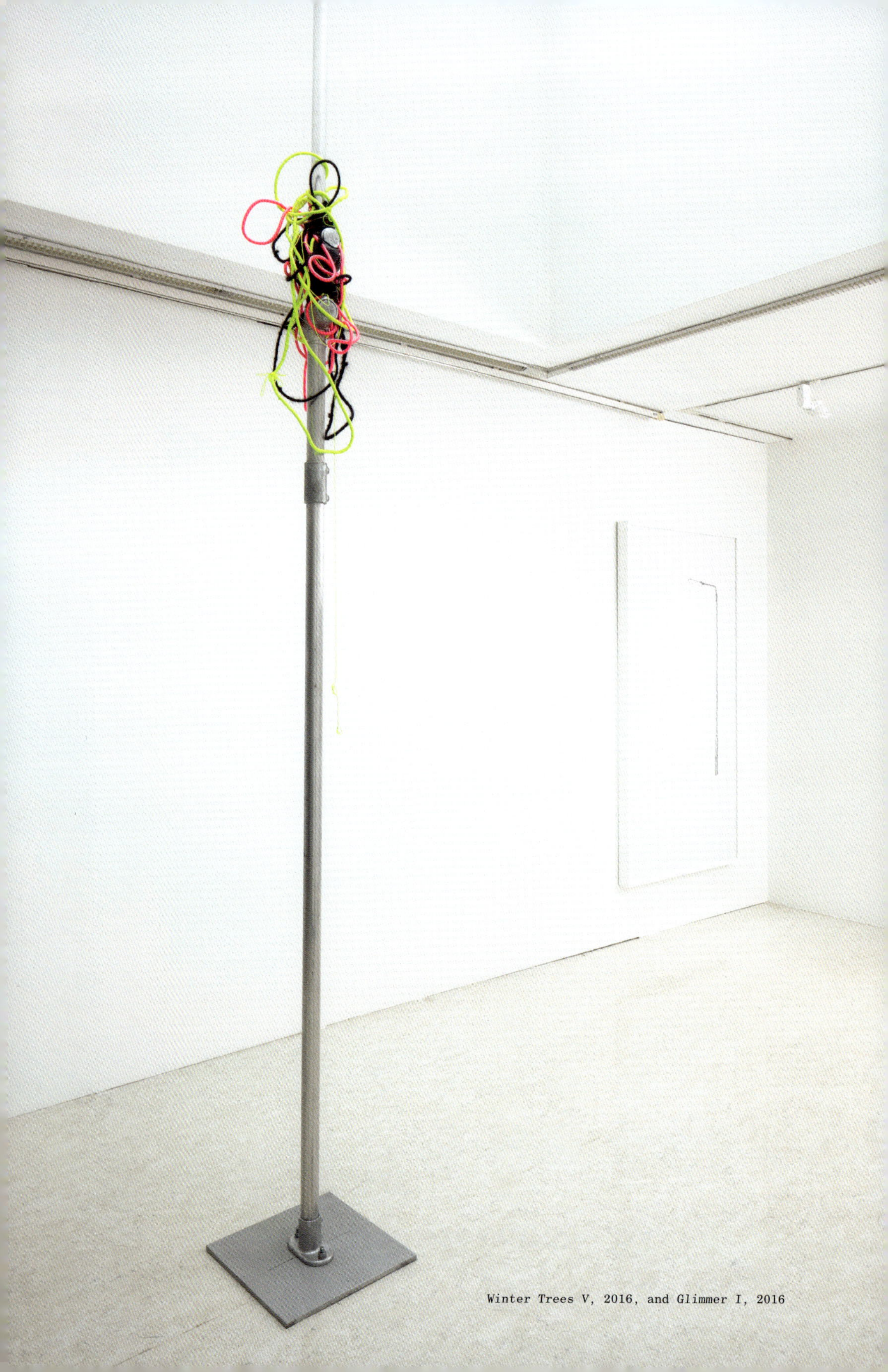

Winter Trees V, 2016, and Glimmer I, 2016

Glimmer IV, 2016

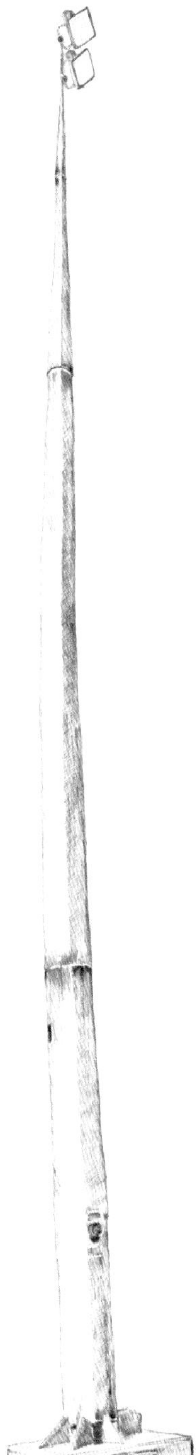

Glimmer V, 2016

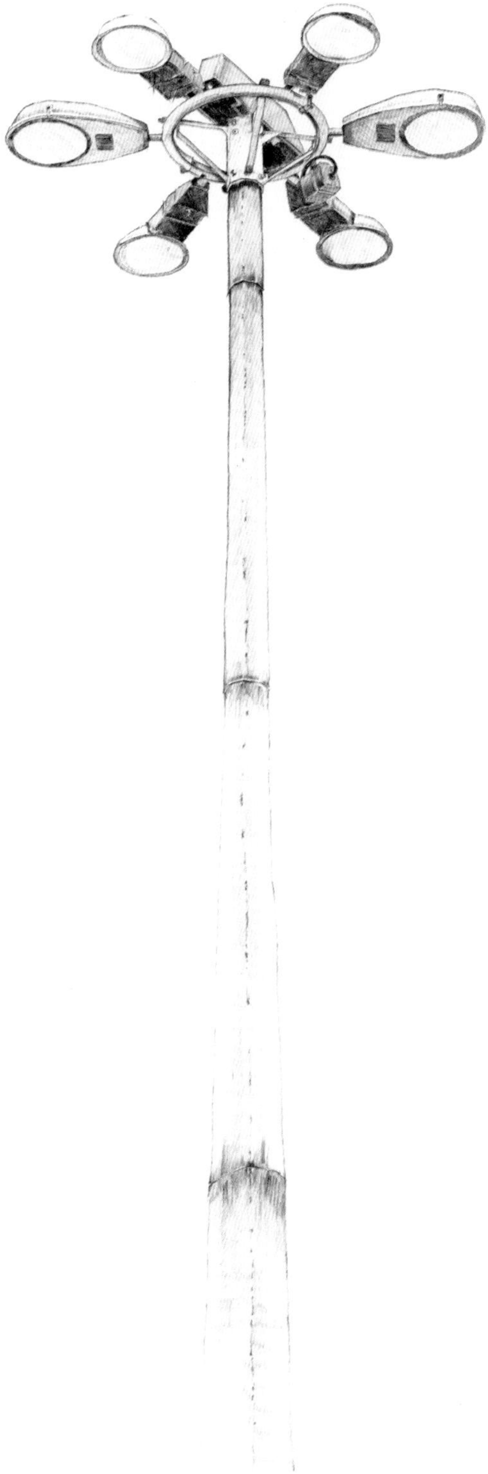

Glimmer II, 2016

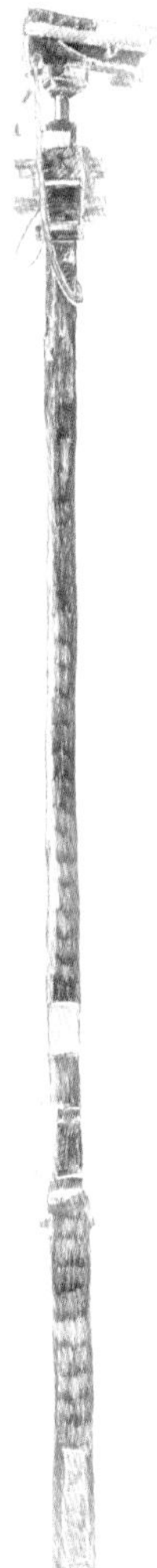

Glimmer III, 2016

Glimmer I, 2016

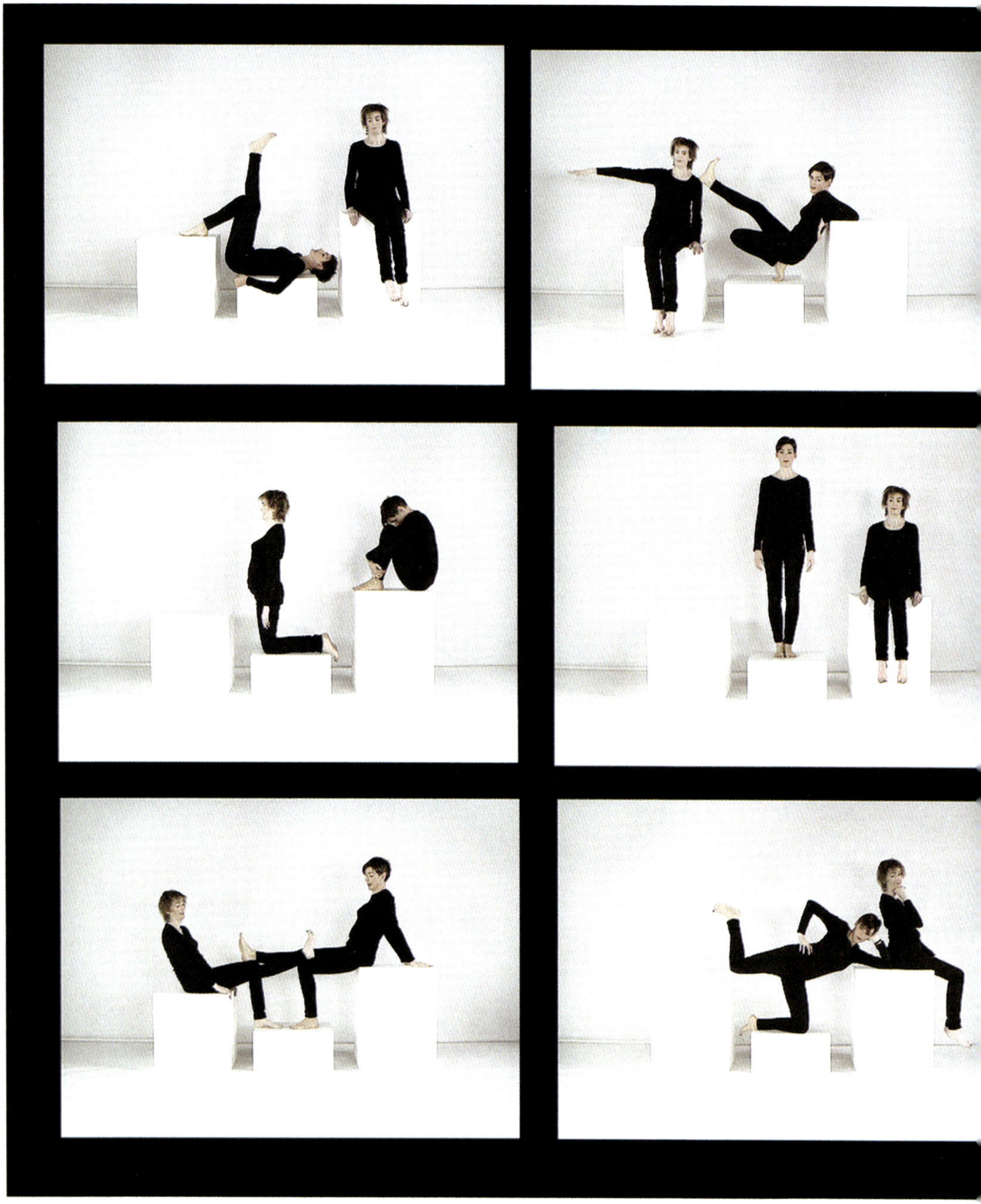

Pose Work for Sisters, 2016 (Glasgow Museums, AV.2017.2)

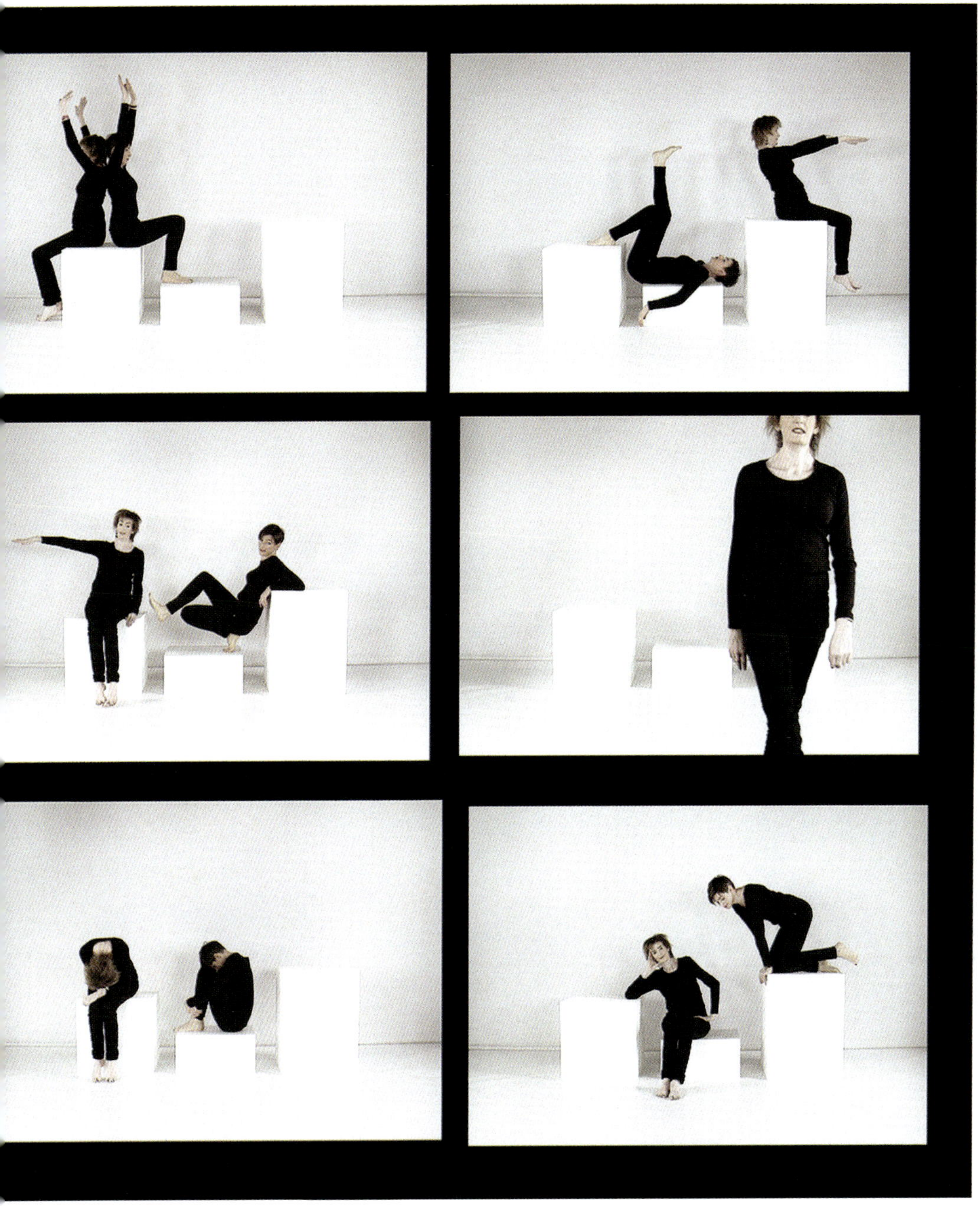

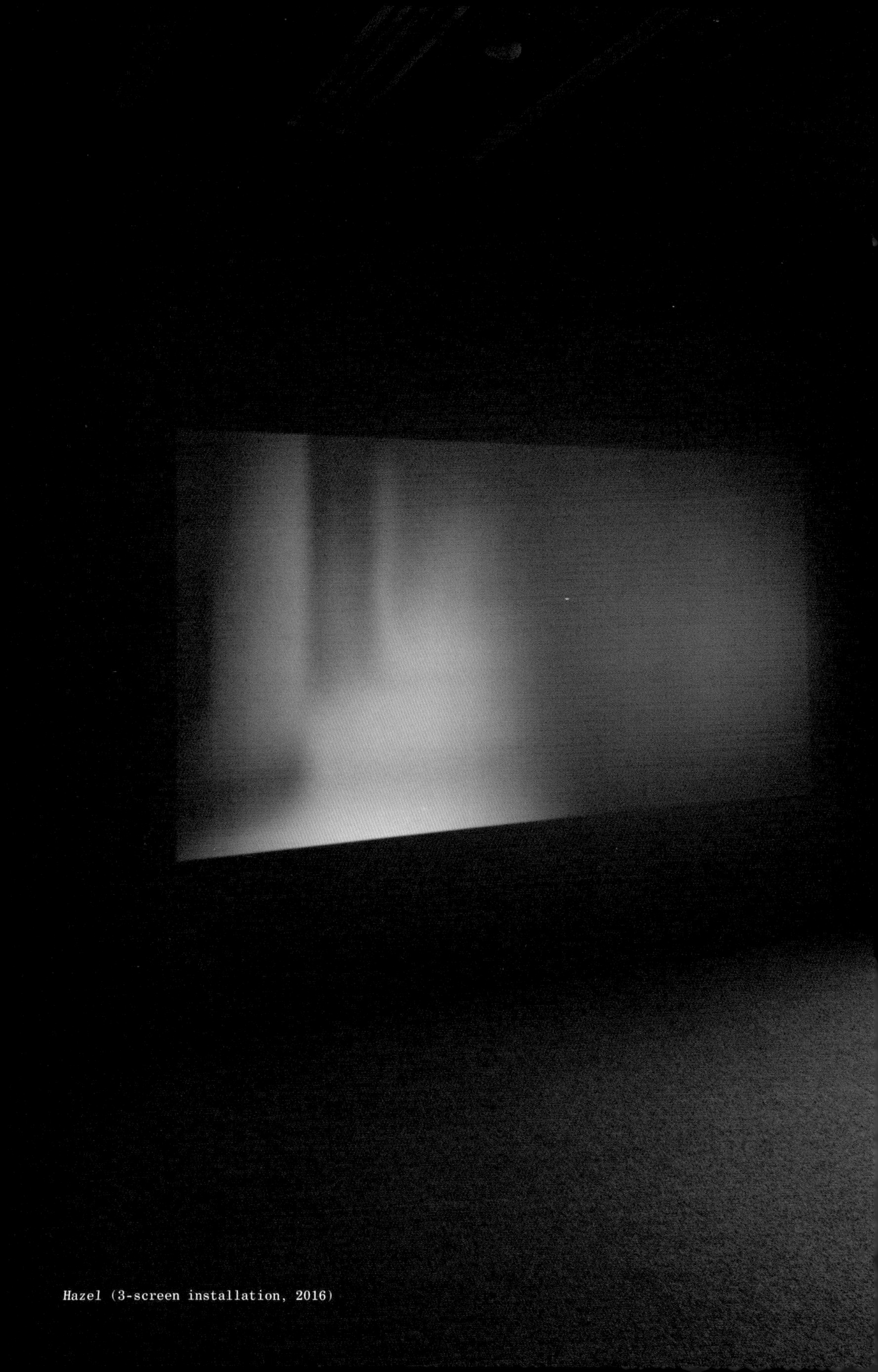

Hazel (3-screen installation, 2016)

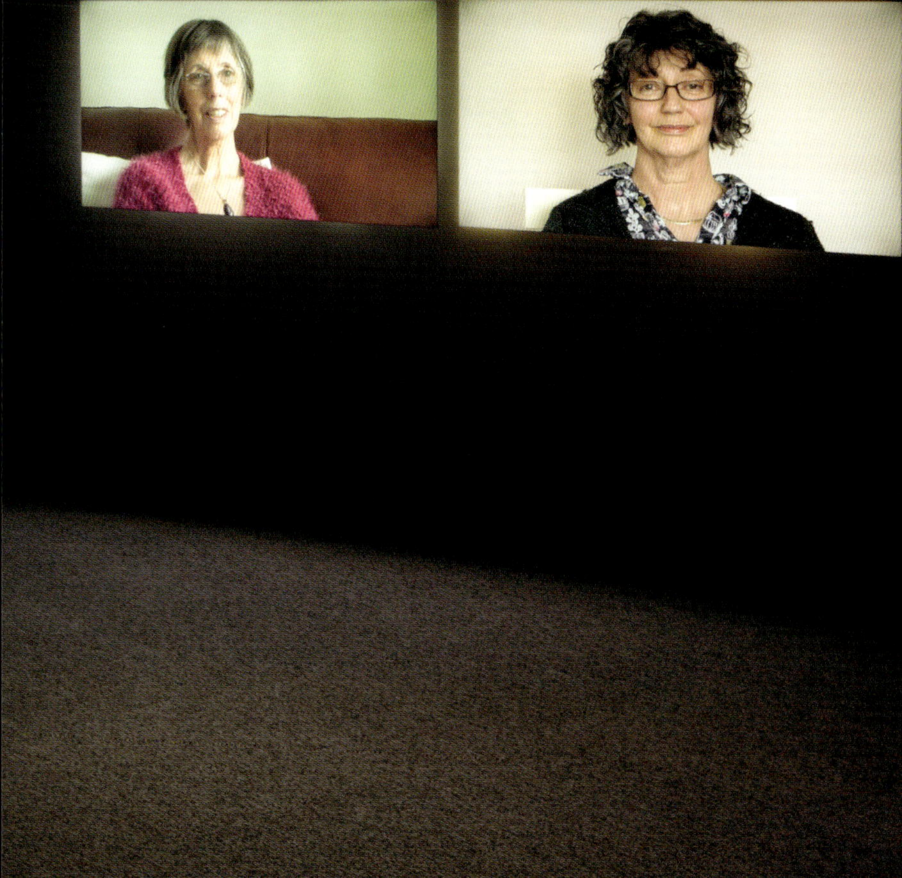

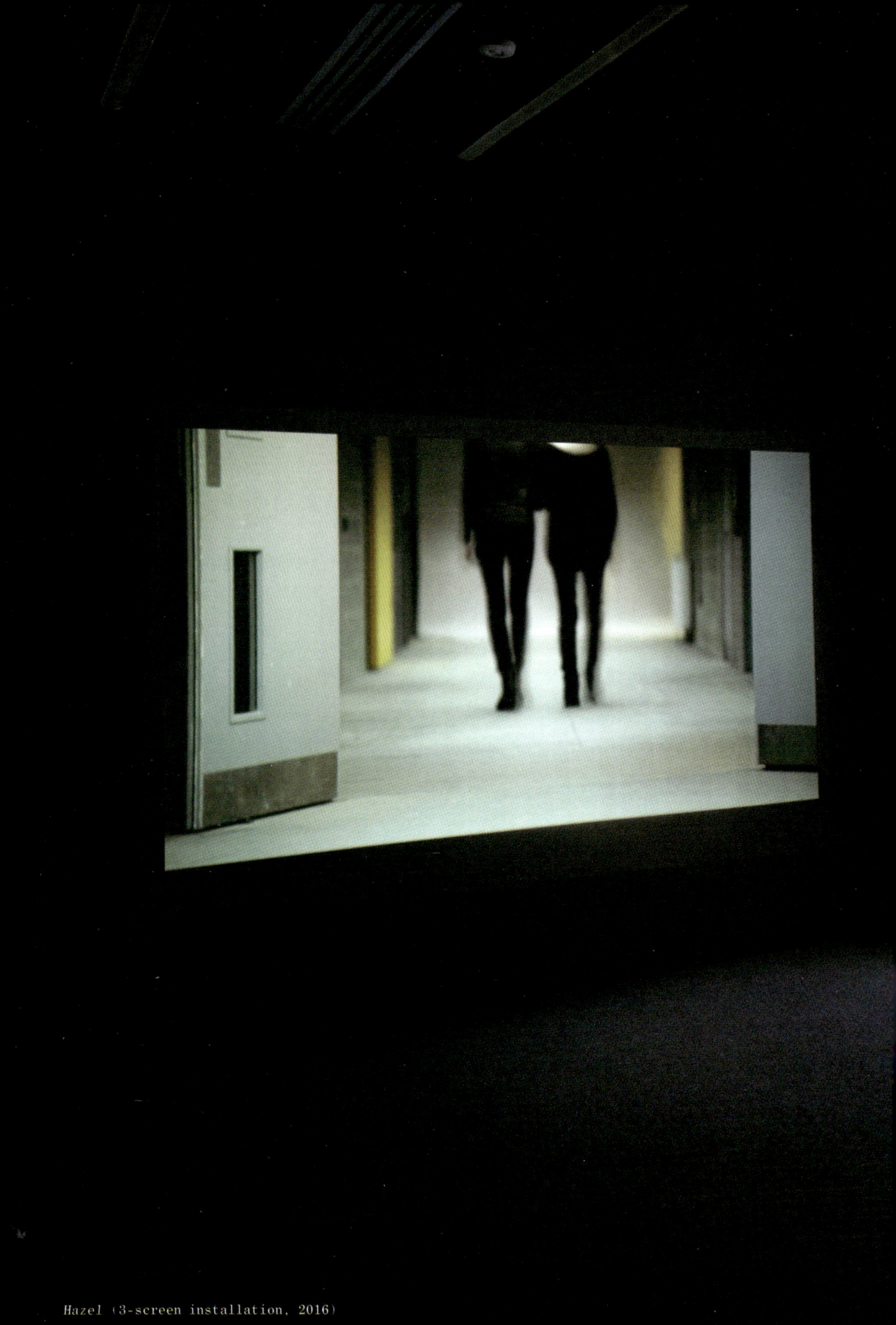

Hazel (3-screen installation, 2016)

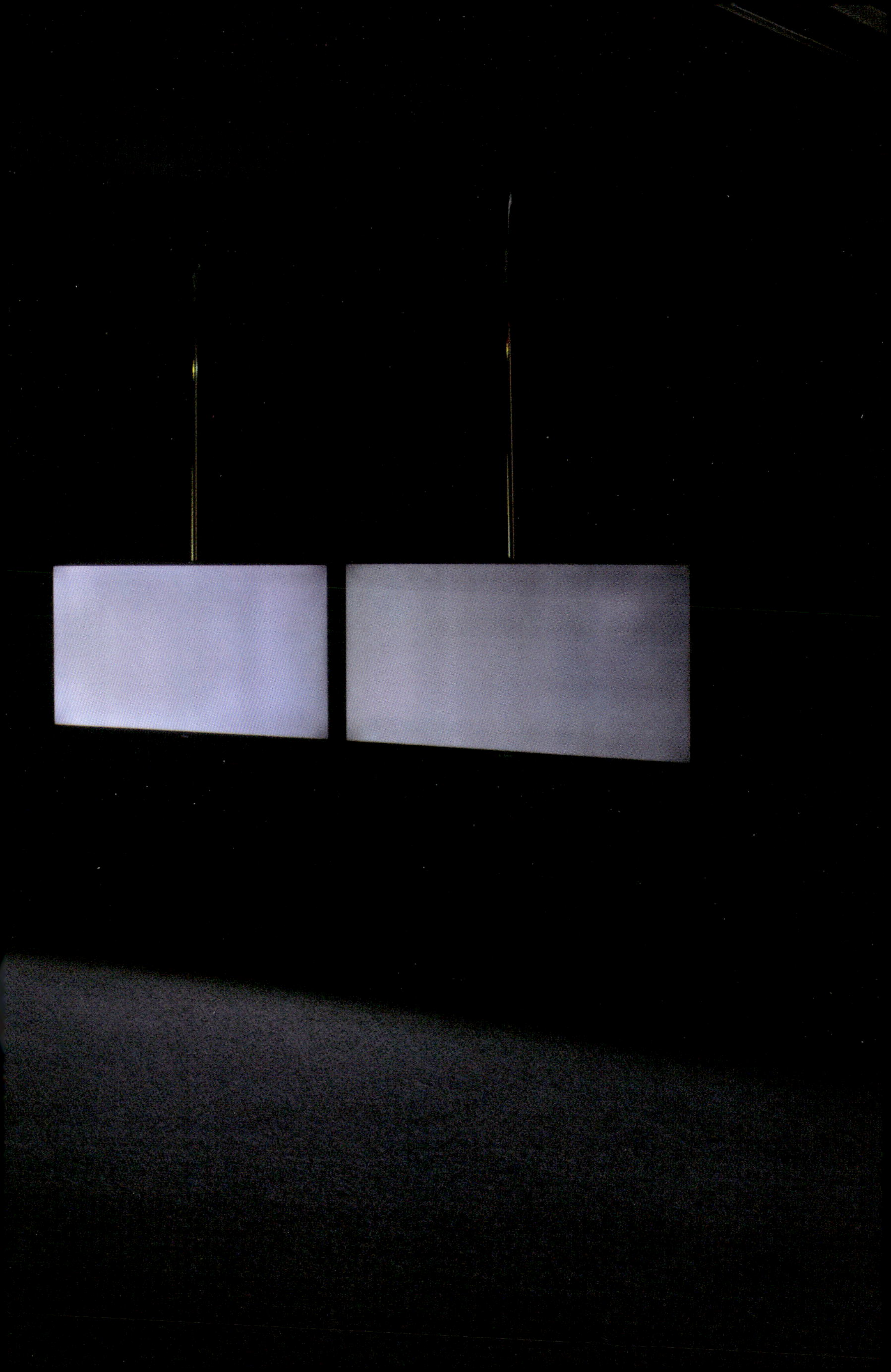

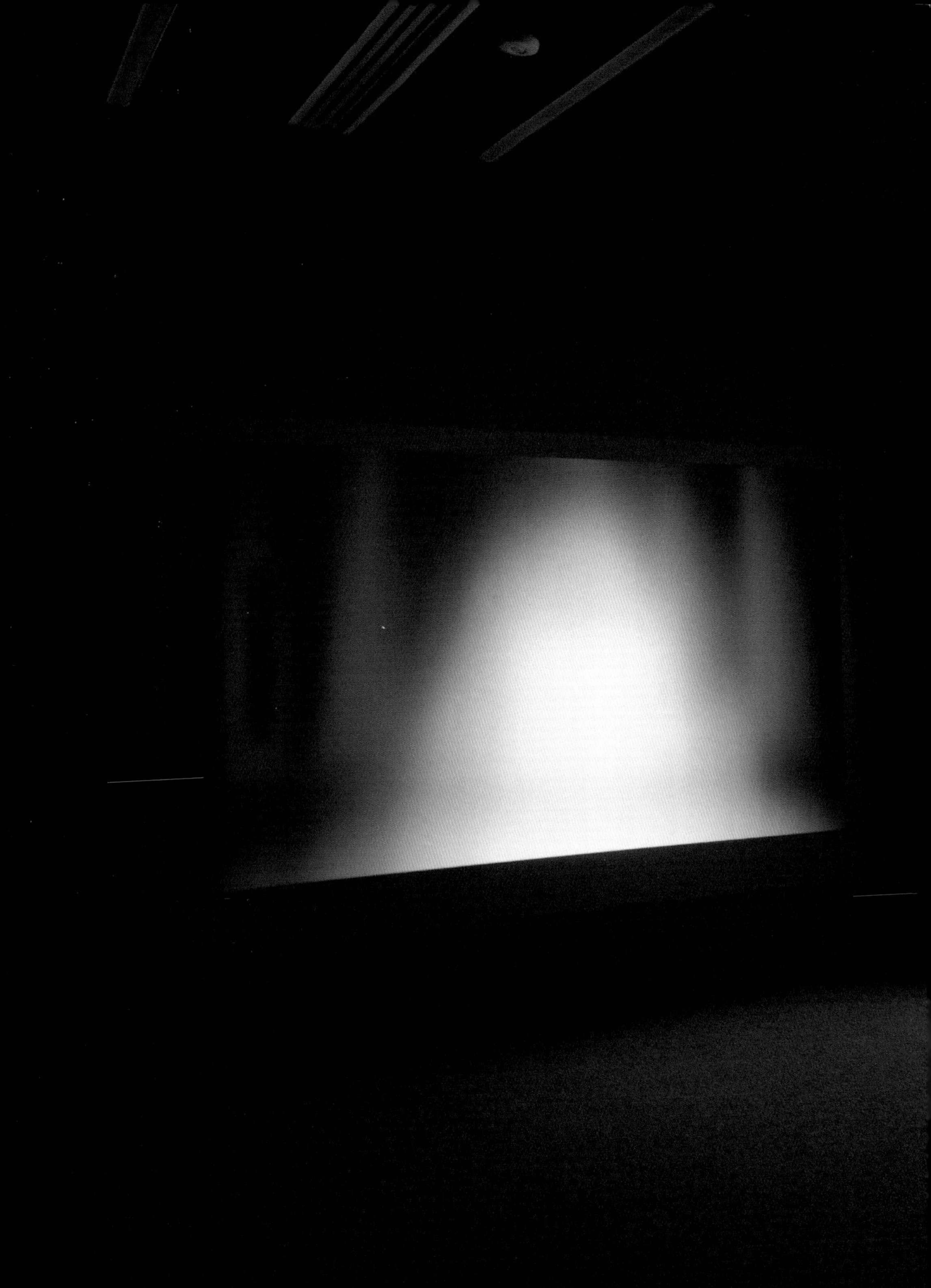

Hazel (3-screen installation, 2016)

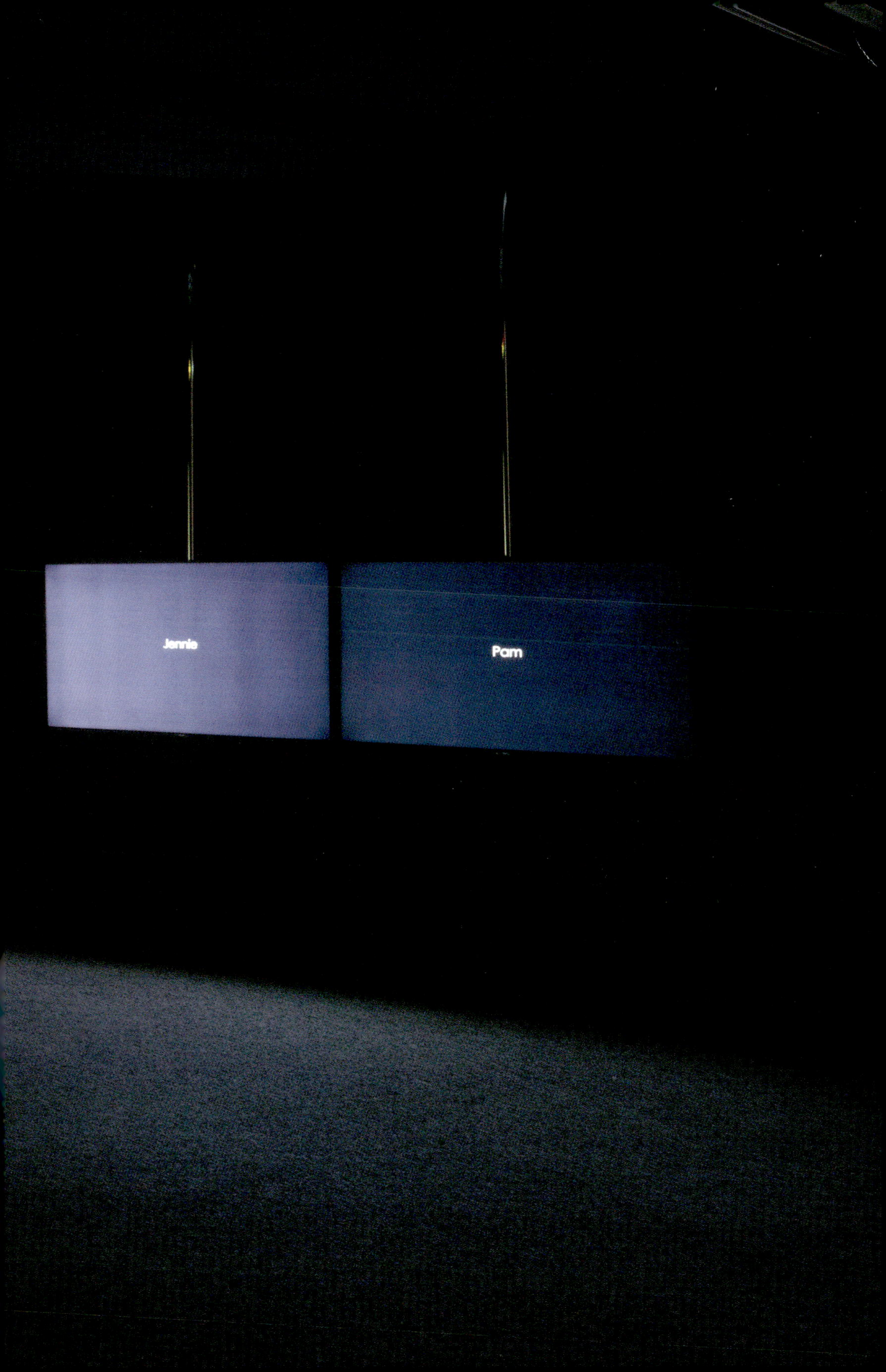
Jennie
Pam

Slingshot

Nicola White

When they built this place in the seventies it must have looked rough, Lauren thinks. Giant concrete and glass shoeboxes in a sloping landscape of bare earth. Mud churned into fat tyre tracks and the wild hills beyond. Sheets of glass dangling from cranes. She had always loved a building site.

And when they'd finished building the university, they planted trees in the raw soil, really gone for it; trees of all kinds and colours. Now, 30 or 40 years on, she guesses, the trees look older than the buildings do. They soften and veil the edges, give the windows something more than sky to reflect. You can't even see the road or the car parks from here. Trees make everything respectable. Plentiful paths snake between the trees, shallow-stepped where they run down or up slopes. You couldn't get a wheelchair down here for love or money. There it is again. Arlene. Now I can go where she can't.

Lauren stubs out her butt on the step beside her. She's not really a smoker, oh no, not like when she was younger. At the doctor's office and on the insurance forms. Non-smoker, tick. Lauren exhales the last lungful, pulls in her stomach to squeeze it all out. Just an emergency rollie. An exception. Times like this, when things needed to be swallowed down, the smoke helps. Just the one. Or two. She will throw the tobacco away later, won't even bring it home.

Check the time. 11.56. Picture of her little girl, Katy, hugging the cat behind the numbers on the screen. She'll be home tomorrow. Her idle hand reaches into her jacket pocket to close around a roll of mints with its tail of tattered paper. Pushes up one white disc with a thumbnail. Can Arlene do that still? Lauren has noticed her sister's hands subtly changing shape.

Lauren jumps up to stand, pivots 180 degrees on one toe. Showing off now. Doing things that Arlene can't. Flicks the mint into the air and catches it in her mouth. Skips up the steps, the very embodiment of wellness. How the chips fall. *It's not my fault.* She tells herself to settle. It's always like this at the DM conferences. And after doctor's appointments or after a day helping Arlene in her house. That time when Keith was away. These urges to run, to jump – the indecent vitality coursing through her lucky body. Such speed, such agility.

Lauren trips on an uneven slab and the word 'agility' dissolves. A stumbling head-forward run, destination earth, but just before she topples, her flung-out arm grabs a railing that her eyes or brain have not consciously noted. How the body saves the body. The handrail is warm from sun, and freshly painted an almost fluorescent red. Perhaps for safety, perhaps someone was bored by the tastefulness of the shady green lawns.

She stops and gathers herself, her body ringing from the almost-fall and the pride that went before it. She holds the metal bar, feels a memory coming down the pipe. Dips a *plié*. Her and Arlene in Mrs Greene's ballet class. She kicks back one pointed foot. *Arabesque. Port de Bras.* Ooof – back spasm. That thick pink leotard, itchy at armpit and crotch. Arlene looking beautiful in her net tutu. Mrs Greene had hopes for Arlene, who she said had the right proportions, but Arlene wasn't that bothered about ballet. *You break my heart, Arlene.*

Lauren is now late for the afternoon session, but hell. She does that thing of pretending she has slipped into Arlene's body as Arlene's body is now. She holds the bar loosely and lets her muscles hang from her skeleton, makes her ankles floppy. Leans on the bar and steps up – one foot, follow with the next foot. One foot, follow with the next foot. Was this how it felt for her sister? Shameful to pretend, yet she aches to know. The entrance to the conference hall comes into view and she lets go of the handrail, walks her normal walk.

Though it's now gone the hour, people are still milling behind the glass wall of the foyer. As Lauren closes in on the building, she has the impression that they are all turning away from her in disapproval, putting down their cups, retreating further in to the building. She speeds up.

Her mother and Arlene are just where she left them, in the back row of the lecture hall. Words come up on the screen – *Patient-Focused Drug Development Strategies* – then it reverts to the blue of a Windows desktop – the arrow cursor speeding around, clicking on folders. Two figures huddle behind the lectern.

Lauren sidesteps along the row and takes her chair. Arlene waves a bag of chocolate eclairs in front of her, shakes it. She dips in and takes two. Unwraps one for her sister and one for herself.

'Don't know if I can be bothered with this,' says Arlene.

Their mother is watching people go up and down the aisle, passing comment in a mutter that is not as subtle as she thinks it is. Two tall, glamorous women pass. One is fair, the other dark, but you can see they too are sisters.

'Do you think one of them has it? They look very well.' One of the tall girls flicks a look back at Lauren and Arlene's mother, her face blank.

Their mother does this at every conference, searches out people who have myotonic dystrophy but look 'well on it'. If someone goes by in a wheelchair she says nothing, just shifts her shoulders slightly, looks at her lap. At coffee breaks and lunches, she befriends. People think her lovely, charming, and indeed she can be, but she is mostly gathering data.

They come to the conferences every year. Lauren comes to demonstrate to Arlene that they are all in this thing together, and sometimes it is interesting, sometimes the experts give you hope that a cure is on the horizon, that they are closing in. Arlene says *she* only comes because their mother wants her to, though last year, in a shared room of a Travelodge, lying in the dark, she told Lauren that the conferences had made her feel less alone, that it wasn't just happening to their family. That's the only time she has said anything that even came close to a complaint. She is remarkable.

At lunchtime, they had talked to a family from Northern Ireland, and Lauren heard Arlene say, *it doesn't really affect me at all.* She sounded sincere, but her wheelchair is parked at the back of the canteen in case of long distances. Lauren's not sure if Arlene is putting a brave face on, or whether she is actually talking about something beyond symptoms, something more fundamental that Lauren can only guess at.

The Northern Irish family were all skinny and dark eyed, mother, father, two sisters, teenage grandson. Just one of the girls with DM and an absent brother who has it too – *he'd rather bide at home.*

When they left the table, Lauren hovered at Arlene's side as they walked off, just in case. The dipping way she walks now. When they are out of earshot Arlene asks – *do I sound like that, Lala?* She was always *Lala* to Arlene, but no one else. A toddler's estimation of her big sister's name.

'No,' Lauren said, 'you definitely sound Scottish', and they had to stop and laugh, snorting, leaning against a corridor wall.

She knew that what Arlene meant was the way the girl slurred when she talked, just a softening of the consonants, a wee squiffy smear to her words.

'You sound fine.'

—

Arlene and Lauren. Their father's choice of names. Perhaps he had a liking only for soft sounds, like someone who could only take their food mushed up. It had made them hard to distinguish, even to their own parents. *Lauralene*! their mother sometimes called. And only a year between the two of them.

When Lauren was ten and Arlene nine, they woke one morning to see barriers being put up on the Links across the road. *The shows!* But there was no sign of the trucks with dodgems on the back or the folded-down chairoplane. Instead there were yellow diggers churning up the grass, waved on by stern men in hardhats. The flat field they had looked out at all their lives, the field where the shows parked up in June, where people ran their dogs or kicked a football was to be turned into houses, their mother explained. (Always their mother. Where was her father? At work perhaps, or lurking in a room they were not in.) There was a big sign on the fence with the words *Executive Homes*.

After the machines stopped for the day, Arlene and Lauren would run across and squeeze through the place where the barrier was loose. It was still their field, they were entitled. It was interesting – the cement floors and low walls, the ditches and piping, pallets of bricks and tiles all wrapped up in shiny plastic. It was a place that Lauren now found herself revisiting in her dreams. She would find herself hiding at night in one of the roofless rooms, or driving a yellow dumper which inevitably started to veer out of control, the detached steering wheel useless in her hands. It's weird what sticks.

When the walls got high enough, their favourite thing was to walk the tops of the new walls, pretending to be on a tightrope or balance beam. They were daring – cartwheeling off heights to land, arms in a triumphant V, like Olympic gymnasts did. That's how Lauren's tooth got broken at the front, and Arlene went and told their mother how it happened and Mum said they weren't to go to the building site anymore. When they got up to their bedroom, she wrestled Arlene to the floor and knelt on her shoulders to pin her. Arlene knew what came next – the silver thread of spit – and thrashed her head side to side, her face all red and screwed up.

All the things Lauren could not undo.

Of course, they defied their mother and went back to the site anyway, the houses rising day by day. The thrill of the upper floors. The drunken feeling of hanging upside down on cross bits of a stud wall, and Arlene doing the same, across the sketched-out room from her, grinning, her long hair hanging straight up, it seemed. How careless they had been with their bodies back then. Her sister's body sometimes seeming like just an extension or a copy of her own, but all the time, there was a difference she could not see, that no one could see. A lengthening of parts of Arlene's DNA, genes with extended repeats, like little tails. That was the way the doctors tried to explain it.

She sort of understood, vaguely, but could only visualize the repeats as hazel catkins – the way they were dense and stubby to start with, then gradually stretching, unravelling, dangling like tiny feather boas. It was the stretching that made the symptoms worse.

One sunny evening, she and Arlene were lying on a stack of breezeblocks, head to head and feet in opposite directions. It must have been a hot week because they were both sunburnt. No one wore sun lotion back then. You wouldn't waste cosmetics on a child. Lauren remembers leaning across and grasping a frilled edge of peeling skin on Arlene's upper arm. She pulled very gently and a big swathe of translucent skin came away. You could see all the little crisscross lines as the sun shone through it. Lauren put the skin in her mouth and chewed it down. Arlene screamed in outrage.

She had loved not being grown up. Muck savages, their mother used to call them. No better than animals. Aside from Lauren's cracked tooth, it was probably Arlene who would get bruised and scraped more often. But that was because they were boisterous, or because Arlene was littler and trying to keep up. But it could have been the DM affecting her agility even then.

That was one of the things that was so frustrating about her sister's disease. It disguised itself in quirks and common complaints. Things too easily dismissed on their own.

It doesn't affect me at all.

Before school started again, the roof beams appeared on top of the breezeblock walls; triangular ribcages waiting for their tiles. Soon there was no more building site, only poky houses.

—

The phone call came in the middle of the night, as bad calls do. When Lauren got to the hospital and asked for Arlene, they sent her up to the sixth floor. Neonatal intensive care, they said, though Lauren didn't really take in what that might mean.

The room was buzzing with nurses, and machines on wheels with see-through lids. But not lids, incubators. Woollen-hatted babies inside, impossibly small. She caught sight of Keith in one corner, but a nurse turned her towards the hand wash station, pump-bottles of red stuff and paper towels and her fingers were shaking as she did her best, wondering all the time where Arlene was.

Keith had a cloth bundle in his arms, and his eyes shone with tears. The baby moved his head slowly towards Lauren and she felt a wash of sudden relief. She had thought Keith was crying, you see, the baby lost, but he was just full of pride.

'Do you want to hold him?'

And to be honest, she didn't, he looked too fragile, and there were tubes taped to him, but Keith meant it as an honour, so she made a cradle of her arms and let him put the baby into them. So light, he was.

'We're going to call him Jack.'

Jack looked too delicate to stay in the world. His eyes were only half open, and looked drugged and far away. He reminded her of a fledgling somehow, open-mouthed and thin. He didn't work his limbs as she'd seen other new babies do, just lay in her arms, like he was exhausted. He had come into the world early – 28 weeks. It was not too early to survive, Keith explained, not at all. She didn't know too much about babies, having none of her own back then.

Arlene was in a ward downstairs. She had reacted badly to the anaesthetic for the emergency Caesarean, and was being observed. Keith sat in the room with the incubators until dawn, and Lauren sat with Arlene who wouldn't, or couldn't, wake. Days blurred into other days at the hospital. Days divided by sandwiches and expensive Costa coffees. Arlene rallied, but was still grey and tired, worried over Jack. He didn't thrive. *Floppy* was the word they kept using, though it didn't sound medical enough to Lauren's ears.

'He's just not helping himself,' a nurse said.

'Well, you help him, then,' Lauren snapped.

Up until Jack's birth she had been feeling a little jealous of Arlene's pregnancy, but all that was forgotten now.

The hospital couldn't work out what was wrong with Jack at first. Eventually they tested him for myotonic dystrophy, a condition they had never heard of. Arlene claimed they hadn't told her they were doing the test. Maybe they did, though, Lauren thought. It was so hard to hang on to all the stuff they said and what it meant.

A few days later, she found herself in the consultant's office with Arlene. She had thought she was there in a supportive role, but then he said they should both take the test. They drew blood, then sent them to talk to a genetic counsellor, a perky older woman who asked about their family history, was there anyone who was affected by muscle weakness or muscle stiffness, difficulty walking or gripping, cataracts, diabetes, IBS, lethargy, sensitivity to anaesthetic, premature labour. She was just throwing out all these symptoms like someone emptying an attic. Lauren was overwhelmed by a sense of oncoming misfortune.

'Has anyone among your relations given birth to children who didn't survive infancy?'

Arlene made a little gasping noise, and the woman frowned at her in concern, then remembered and apologized.

'I'm not implying anything about *your* child,' she said, and pushed a box of tissues over the desk to Lauren, who was surprised to find she was weeping.

'Our father doesn't keep great health,' Arlene said.

'But he's okay,' Lauren clarified, reaching for a second tissue. 'I mean, he's getting on, of course, he's in his late fifties.'

'He has cataracts.' said Arlene. In her memory, Lauren remembers Arlene's voice being clear as a bell, and followed by too long a silence, as the counsellor writes a hurried note.

It seemed that little Jack had lit a fuse that was running and branching throughout their family. No, that was wrong, for the fuse had been lit elsewhere, far back in a dead generation, possibly, and it was Jack that was at the end of it. They remembered Auntie Sally, their dad's sister, who was painfully thin and walked with a stick, then their dad's brother Eugene who died as a baby, they didn't know how old. Grandad

Dunn died young too, but they thought that was only a heart attack. Granny Dunn had trouble with her eyes, but was in her eighties now.

The counsellor scribbled all the while, and her eyes sharpened and flashed with every new connection. For her, something was coming together. For Lauren, it was the opposite.

Then their results came back.

Arlene did have it, and had passed it to Jack. They confirmed that the disorder had probably been passed down through their father. Further tests would be necessary for their wider family.

Lauren was in the clear. *You are unaffected,* was what the counsellor said.

Her nephew lived for two months and 11 days.

—

After the conference, Lauren drops her mother off first, promises to call by at the weekend to help with Dad, then drives Arlene to her house. Arlene hums a tune in the passenger seat.

'Do you want the radio on?' Lauren asks.

'No, it's fine.'

'So was it interesting for you, the conference?'

'I guess. Those people from – what was it – Larne? – they were really nice.'

'When they were talking about what drugs could do for symptoms – was there anything sounded helpful?'

Arlene shrugs, picks a CD out of the door pocket. She can be hard to talk to about the illness.

'I mean … if you could take one of your symptoms away, which one would you get rid of? If you had a magic wand.'

'You know I don't like those games.'

'It's not a game. I'm interested in what you think.'

'I know, but there's no point. There's no wand.'

'I'm not trying to annoy you.'

They are almost at Arlene's house. Lauren turns left into the street. They don't speak again until the car is parked, the engine quiet.

'I'm fine, Lala. You worry too much.'

Someone has to worry, Lauren thinks. She walks around to open the car door for her sister, then goes to get the wheelchair out of the boot. The streetlamps flicker on suddenly, dimming the rest of the world into twilight. Arlene is standing half way up her garden path, looking down at the ground. She must be tired after the weekend.

But when Lauren gets near her, she sees she is looking at something in the flowerbed, pointing.

'Look, the Queen of the Night has come out.'

In a clump of pointed green tulip buds, a silky black cup has emerged, purple where the streetlight hits its curve.

'It's soooo beautiful,' Arlene croons. 'And look at the columbine, every year they come back. I don't have to do anything.' She touches a pale blue flower gently with her stick.

Lauren can't manoeuvre the wheelchair around her sister and her sister's walking stick, so she waits, determined to be patient. She notices the sound of running water and looks over at the little fountain that Keith and Arlene have put in the garden in memory of Jack. She could do without the cherub posing on the top, they never did have the same taste in things, but the idea is lovely. Arlene should have another baby. They have methods now to ensure the affected genes won't be passed on, though it wouldn't be easy. Difficult road and all that, but possible.

'Arlene? Have you and Keith thought more about – '

'Shhhhhh…' says Arlene, smiling at her flowers.

Behind the lit front window, Keith appears, waves, does a little knife and fork mime to signal that food is waiting.

—

They would stay at the building site up until dark that summer. There were big metal plates that spanned the deep ditches where the plastic pipes were laid. These were for the heavy machines to drive over. They were good for bouncing on. Not as good as a trampoline, but they had flex, and the booming sound when you landed back down was thrilling. They would jump and jump.

Surely they had both been as strong as the other then, as fit. Lauren wanted to believe that there had been a time when there was no difference between them. When the mesh of their DNA was equally resilient. She doesn't know why it matters to think this.

She suddenly remembers her sister resting on the dusty ground while she bounced on, chanting *Arlene Dunn's a lazy bastard* to the tune of 'Dusty Bluebells'. Was it the condition, or had her sister just been ordinarily tired? It was impossible to pin it down, so why did she keep trying? It was everything and nothing.

Lauren stands by her car in the dark outside her sister's house and remembers hard, under the glimmer of the streetlights. Her sister's curtains are closed now, their living room cosy and unknowable.

They used to run home hand in hand, so they could play what they called *slingshot*. To do it, they picked up as much speed as they could, then one of them – well, usually Lauren – would grab at a pole or a lamp post and swing around it, wheeching her sister past her – fast, faster.

She can almost hear that *Lauralene!* ringing in the air as their mother called them home, and the wrench of their clasped hands breaking apart, as Arlene goes reeling, spinning ahead, untethered, towards the silhouette of Mum at their front door, towards teatime, towards everything that followed.

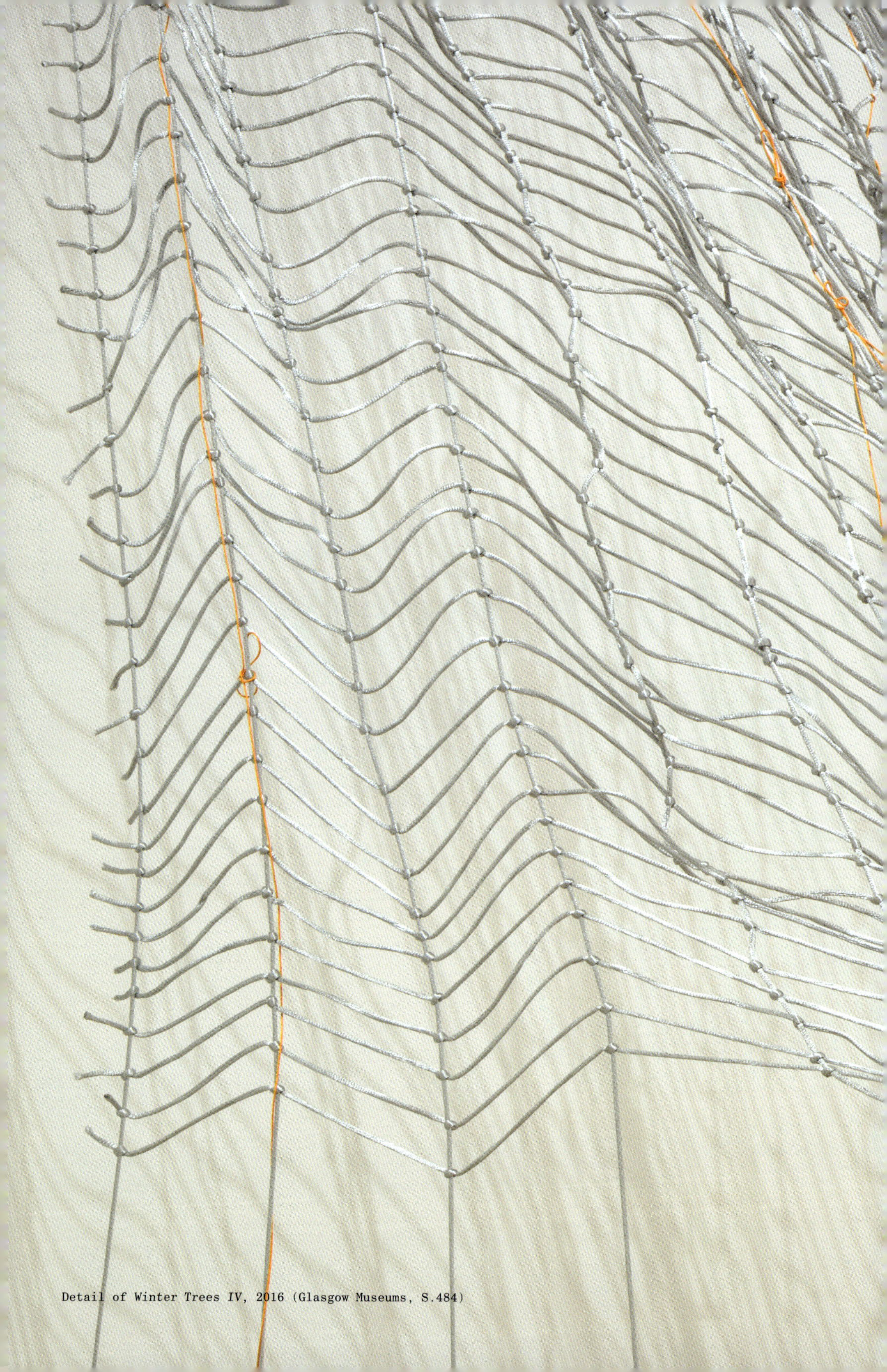

Detail of *Winter Trees IV*, 2016 (Glasgow Museums, S.484)

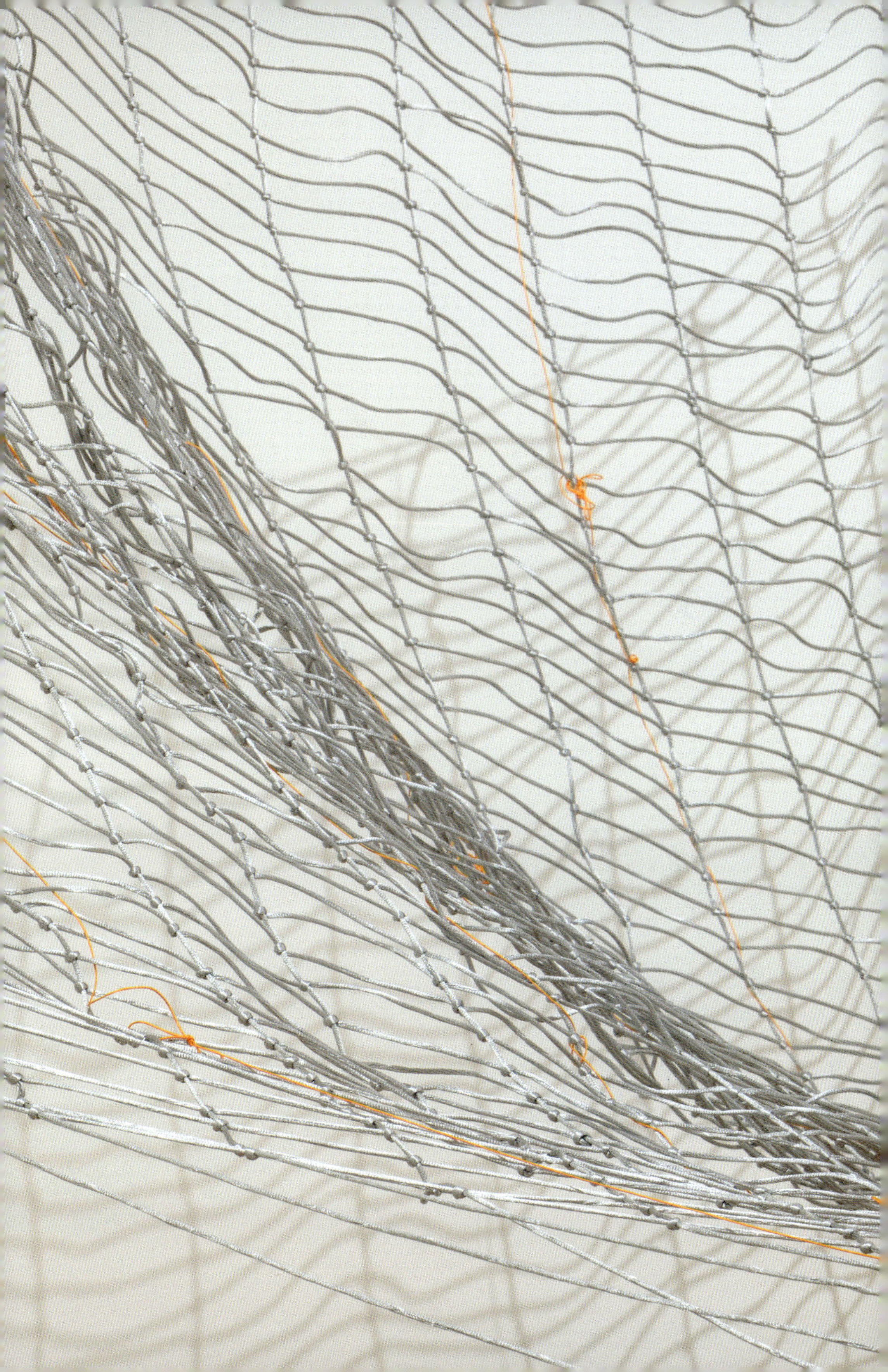

Field Notes from the Heart's Frontier

Moira Jeffrey

Some responses to the *Deep in the Heart of Your Brain* symposium at Platform, Glasgow, November 4, 2016.

Autobiography

In Jacqueline Donachie's film installation *Hazel* (2016), two screens are suspended side by side in darkness. On one screen, we see a succession of five women in domestic settings who sit silently for the camera. As we watch these women they look back towards us, breathe, and occasionally swallow. It is hard to escape the suggestion that they have been silenced. On the other screen, we see a parallel series of five talking heads. These women talk calmly, frankly and openly, telling stories of frustration and, at times, of unfathomable loss. They speak of disability, of changing bodies, and of changed lives.

All the women, who address the camera directly in speech or silently in gaze, are identified only by their first names. We slowly realize how the sometimes dissimilar appearances of each pair of women are deceptive. Beyond the immediate registers of facial similarity or difference, they share deeper characteristics such as eye colour or something indefinable that may be an almost imperceptible gesture. It slowly dawns on the viewer that the women are sister pairs. But if we can sense a trace of shared ancestry, we also note that all of the women whose voices we hear, themselves share a further resemblance that is almost familial in itself. Both sets of women are dealing in their own private ways with the knowledge of the differences and the similarities that the film has rendered visible. The talking sisters have each inherited the gene for myotonic dystrophy, a progressive neuro-muscular disorder. The silent sisters have not.

Deep in the Heart of Your Brain, the solo exhibition at the Gallery of Modern Art, Glasgow, in which *Hazel* was first shown as a gallery installation, was a show about family, about ageing, about love and lineage. It was about the best of circumstances and some of the hardest. If *Hazel* is a filmed portrait of sisterhood, so too were the tall drawings of urban streetlights that hung on the gallery walls, each bearing a family resemblance to the others. *Winter Trees III,* the sculpture of a metal pole clad in a tangle of neon rope that greeted the viewer at the entrance was, the artist has explained, a fleeting memory of Donachie and her own sister out dancing. The suspended leg of a suit of medieval armour, that hung from the ceiling in the gallery, was both an image of physical restriction and an urgent injunction never to mistake the bodily casing for the person inside.

A low box-shaped sculpture entitled *Deep in the Heart of Your Brain is a Lever* was made by altering and powder-coating the metal sheets that are conventionally used to make vehicular ramps. An orange handrail ran through the space, making awkward jumps of height and angle. Entitled *Nice Style,* it suggested a route through city spaces or the way, Donachie recalls, her father used a washing line to support himself as he walked. But it also told of the mysterious thread of genetics, of family history, of shared endeavour and care.

As the short sequence of talking heads in *Hazel* comes to an end a third screen flickers into life. Two figures walk side by side down a long corridor. Their similarities – they are both tall, rangy, dark-haired and dressed in black, their arms hanging loosely by their side – are offset by minor but distinct differences in their mobility and gait. The stories in the film mirror Donachie's own life. This is the story of the artist and her sister.

Epiphany

The roots of autobiography in western literature lie in stories of spiritual pilgrimage and conversion. Life writing begins as the tales of apostles and saints. Stories, at first, of the trials of great men. Then, later, of the immeasurable suffering of women.

A life only becomes a life story, that is, a matter of literature or a matter of art, in the light of revelation and personal transformation. The life story is not just a representation of transformation but is itself an agent of transformation. As an art form, it emerges not to delineate the truth of its subjects' suffering, but to turn the messy, uneven contours of the lived life into the well-defined shape of the instructional example.

At the heart of the autobiography is the moment of epiphany, once a moment of religious revelation, now the manifestation of meaning: a discovery, an illumination. The epiphanic way of thinking and writing has persisted for centuries in the way we shape secular life stories. In scientific biographies, it persists as the moment of scientific discovery. In romantic poetry, it is Wordsworth and his daffodils. In modernist literature, it is James Joyce's alter ego Stephen Daedalus and the momentary lapses that reveal his inner truth. For in the twentieth century, the century of the self, an epiphany became a stripping back, the laying bare of authenticity, often through errors, slippage or inadvertent gesture.

But what if epiphany offers neither transformation nor compensation? Neither religious revelation nor self-knowledge. Researchers working with people with life-limiting conditions talk about the 'negative epiphany' of diagnosis. In progressive or inherited conditions, diagnosis may be a slow realization, an agonizing wait in myriad consulting rooms. It may take years of unexplained ill health or generations of unexplained loss before a diagnostic conclusion is reached. And what if that conclusion is life-altering in unimagined ways?

What if our life story is not a series of transformative epiphanies but instead that slow walk down a concrete-floored corridor? How might we write or speak or show the lived life? How, despite difference, do we remain beside our sisters?

Lightbulb

A cartoon epiphany is easily visually represented. It is the lightbulb moment. Donachie calls her recent body of work on myotonic dystrophy, and her doctoral research, 'the illumination of loss'. The overhead lightbulb eliminates darkness through the even distribution of its value. The torch, which moves selectively, tells us that beyond its bright perimeter the darkness is still there.

Installation view of *Headphone Music, Boats and Trains*, 2016, and *Nice Style*, 2016 (Glasgow Museums, S.482)

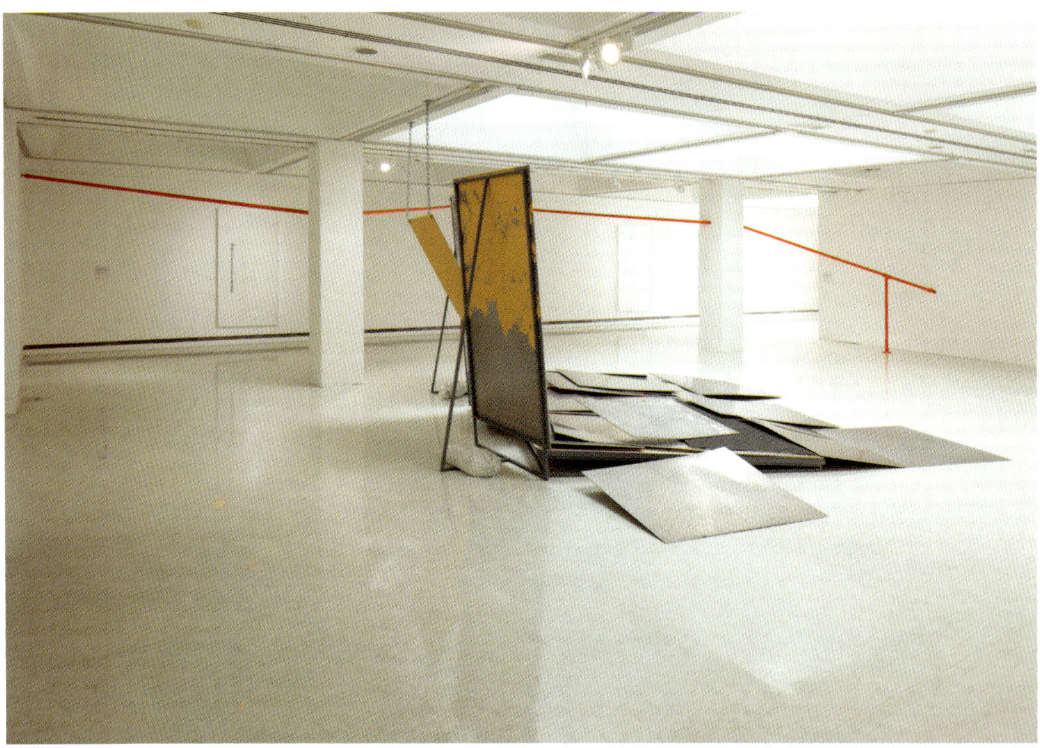

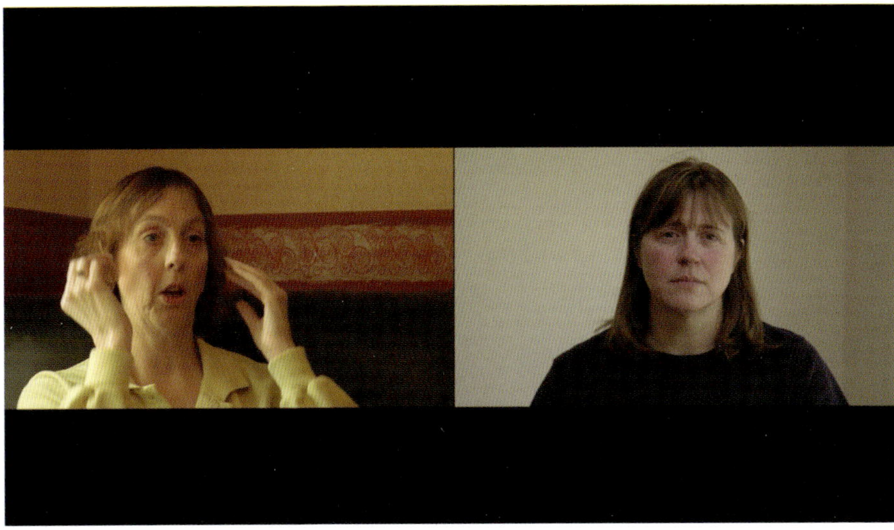

Hazel, 2015

Manifestation

The very idea of epiphany is dependent on a visual manifestation: a display, a reveal, or a striking appearance. But *Hazel* suggests that the very visibility of disability is a complex and uneven matter. When Donachie invited medical experts to see the film installation at the Gallery of Modern Art in Glasgow, she asked them if they saw the family resemblance between the sister pairs. 'They could only see the myotonic dystrophy running down the left-hand side.'

Scientists deal in cells and samples. They read the body in extreme close-up. Physicians are taught to read the body diagnostically. Certain types of scientific gaze may seek data even in intimacy. Think of Charles Darwin noting the facial expressions of his son William Erasmus born in December 1839. Writing in his diary a series of scientific notes rather than of the joy of fatherhood. 'During first week', Darwin writes, 'yawned, streatched [sic] himself just like old person – chiefly upper extremities – hiccupped – sneezes sucked…' This doesn't mean he didn't love his tiny son, but that he observed him as well as saw him. That he saw in him what others did not or could not.

A lay person reads *Hazel* in a different way from an expert. I might read a face searching for emotional expression, family resemblance, reassurance or meaning. In the faces of the silent sisters, I can read everyday progressions: the different pathways of shared genetic heritage, the impact of ageing. In *Hazel*, those of us with no direct experience of the condition learn that the decreased facial mobility of patients with myotonic dystrophy gives the mouth a downward cast. A smile is a straight line, not an upward tilt, explains Louise, while smiling. How do we read others? How differently do others read our own faces and bodies from the way we read our own?

Can we recognize what others see in us? This is not a diagnostic question that refers to disability, but a wider question about the work in progress of being human.

Data

Hazel is one of a group of newly commissioned works developed from a period of research with a group of women affected by an inherited genetic condition made in a unique collaboration with the UK Myotonic Dystrophy Patient Registry, at The John Walton Muscular Dystrophy Research Centre in Newcastle, England.

Registries exist to collect or organize standardized information about particular conditions or diseases. They can be used to learn about a condition, to develop research, for improving or monitoring health care, and for studying best practice. A registry may pursue a specific research agenda, or may collect data on an indefinite basis to answer existing or emerging research questions.

The American poet and artist Anne Boyer, who was treated for cancer in 2014, has written vividly of the dynamics of illness and care. 'The word *care* rarely calls to mind a keyboard,' she writes. 'The work, often unwaged or poorly paid, of those who perform care, (or what is sometimes called "reproductive labour," reproducing oneself and others as living bodies each day, of feeding, cleaning, tending, and so on) is what many understand to be that which is the least technological, the most affective and intuitive. *Care* is so often understood as a mode of feeling, neighbouring, as it does, to love. Care seems as removed from quantification as the cared-for person's sensations of weakness or pain seem removed from statistics class…'[1]

But when she is ill Boyer observes that the work of care necessarily involves the gathering of data. And like other forms of caring, it is often women's work. The mother who fills in the referral form for her child. The nurse who takes blood samples, temperature readings or monitors blood pressure. The collection of information for insurance forms or social security. The experience of illness, and the act of caring, transform the suffering body not just into the object of medical or scientific scrutiny, but into numbers and diagrams.

Data suggests a form of biography or autobiography that is vital but unreadable. Might the artist's job, when dealing with the lived life, be to turn data back into the body again? To translate the statistic back into the lived experience. Donachie's verbatim recordings of women with myotonic dystrophy extend knowledge of the condition's social and emotional impact far beyond the usual clinical data. However Donachie is clear about the nature of her labour. 'Working with research the end result is not information,' she says. 'The end result is art.'

Installation view of *Nice Style*, 2016 (Glasgow Museums, S.482), *Headphone Music, Boats and Trains*, 2016, and *Deep in the Heart of Your Brain is a Lever*, 2014 (Glasgow Museums, S.483)

Silence

The expert's need for data means that the expert cannot deal with silence. One professional response to *Hazel* has been that the silence of the unaffected sisters feels like a missed opportunity for clinicians. Yet the lived life is full of silences, of things felt or not said. Silences that are accidental or enforced. Silences that are willed or purposeful. In active silence we might find the body's resistance against the narrative. In silence we may allow new narratives to flourish. Is silence one weapon against data's empire? Notably, Donachie silences one set of sisters so that another set might be heard. This is a reversal of the conventional hierarchies of silence.

Time and narrative

In ill health and in disability time runs differently. 'Time,' says Anne Boyer, 'acquires a compelling texture for the sick.' Sociologists of medicine have long identified that a significant impact of ill health or disability is what is described as biographical disruption. It is not just a crisis or a disruptive life event, but a change in the meaning one ascribes to one's life story, a change in the narrative arc.

The radical disruption experienced in illness, particularly in disability or in inherited conditions, is not just a matter of an individual's life story being disrupted, but also a fundamental re-ordering of chronologies. In *Hazel*, Louise describes being old before her time, how she feels her chronological age is only just catching up with her impaired mobility.

These chronological reversals reverberate throughout family networks. The re-ordering of time is also the upheaval of expected relationships of reciprocity and mutual care. Biographical disruption can mean that grandparents may become carers of their grandchildren; parents care for adult children; unaffected children may care for their parents. Biographical disruption changes not just the narrative but who gets to be the narrator.

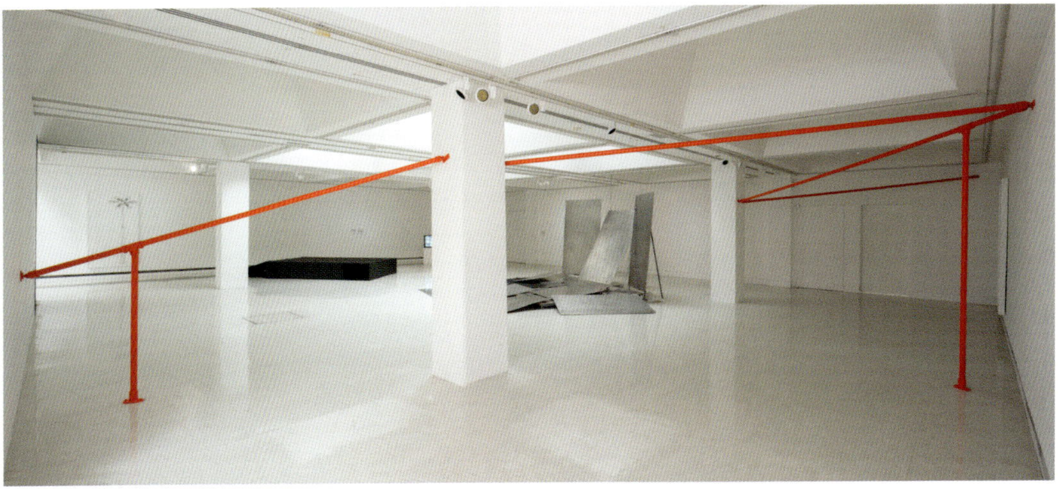

Autoethnography

How can we work with lived experiences to make meanings, which are not pre-packaged as information, revelation or transformation? Autoethnography is one route, fiction another.

Hazel begins with an uncertain image of light and darkness. In its final scene, the focus resolves to show an image of the artist and her sister. In traditional ethnographic studies, researchers conduct field work: they produce what are described as 'thick descriptions' of a culture through techniques like observation, interview, field notes and artefacts. In autoethnography, researchers are participant observers, and these thick descriptions relate to personal and interpersonal experience.

'The autoethnographer not only tries to make personal experience meaningful and cultural experience engaging but also, by producing accessible texts, she or he may be able to reach wider and more diverse mass audiences that traditional research usually disregards, a move that can make personal and social change possible for more people.'[2]

If the artist acts as autoethnographer how might we understand the result? Where are the boundaries of scholarship on this troubled border with authorship? Who is the subject of the artist's autoethnography? And who holds power?

Fiction

The artist shapes data into art, the writer turns it into literature. What is the nature of this transaction? When the narrative arc changes, we change not only the stories that we tell others, but also the stories we tell ourselves.

'Fiction is no longer seen as "false" or "lies" or "make-believe",' writes the critic Jonathan Sturgeon. 'Instead it is more like Kenneth Burke's definition of literature as "equipment for living". Fiction includes the narratives we tell ourselves, and the stories we're told, on the path between birth and death.'[3]

Care

Hazel is a story about calamity that is told *sotto voce*, in calm pragmatism and sometimes through silence. The act of care is implied in the domestic setting. In *Hazel*, the sibling relationship implies love. Are our desires to equate family relationships with love and care justified? And can we recognize the work that love sometimes does without valorizing its sacrifices?

'An economy based on love ends up being an economy of exhaustion – after all, love is utterly exhausting – of deregulation, extraction and lawlessness,' writes the artist Hito Steyerl. 'All is fair in love and war. It doesn't mean that love isn't true or passionate, but just that love is usually uneven, utterly unfair and asymmetric, just as capital tends to be distributed nowadays. It would be great to have a little bit less love, a little more infrastructure.'[4]

By working with the economy of love, the economy of the family, the politics of care or disability, the autoethnographer transforms the personal into the social. He or she crosses, as the pioneering sociologist Charles Wright Mills would have it, from private troubles to public issues. The discipline of sociology, like the disciplines of art and literature, must manage both a sense of detachment and a sense of engagement. The sociologist must be informed by direct experience but examine the public issues that shape that experience. For Wright Mills, a sociologist makes a distinction between public and private but must constantly iterate that distinction.

'The deep embodied experience of health and illness,' says the medical sociologist Professor Sarah Cunningham Burley, 'is also something we have to understand and confront at different levels... Health is a major public issue, and we know that there are multiple social determinants of health. We must recognize this if we are going to understand health and illness, even in all its agony and fleshiness.'

In *Hazel*, it is the familiar circumstances and cumulative effect of sisterhood, the multiple voices of Donachie's filmed subjects, that help us cross from the private into the public domain. *Hazel* not only gives us the voice of lived experience but stages, in a deliberate fashion, one manifestation of the striking disparities that genetic disability can involve. It is truth. It is art. It is data. It is the body. It is private. It is public. Field notes: from the heart's frontier.

1. Anne Boyer, 'Data's Work is Never Done', *Guernica Mag*, March 13, 2015

2. Carolyn Ellis, Tony E. Adams and Arthur P. Bochner, 'Autoethnography: an Overview', *Forum: Qualitative Social Research*, 12:1, Art.10 – January 2011

3. Jonathan Sturgeon, 'The death of the post-modern novel and the rise of autofiction', flavorwire.com

4. Hito Steyerl, 'Politics of Post-Representation', *DIS Magazine*, dismagazine.com

All other quotes are from the symposium, Platform, Glasgow, November 2016

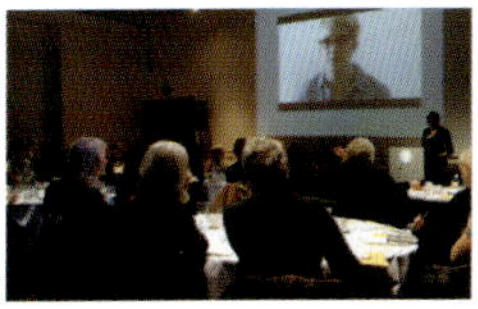

Note:
The *Deep in the Heart of Your Brain* symposium was held at Platform, Glasgow, on November 4, 2016, to develop the themes of Jacqueline Donachie's exhibition of the same name at the Gallery of Modern Art, Glasgow. Its aims were to bring artists, institutions, academics and interested individuals together to discuss care, bravery, lived experience, autoethnography and expert cultures in relation to the ethics and practice of knowledge exchange/ public engagement in the art/medical research field.

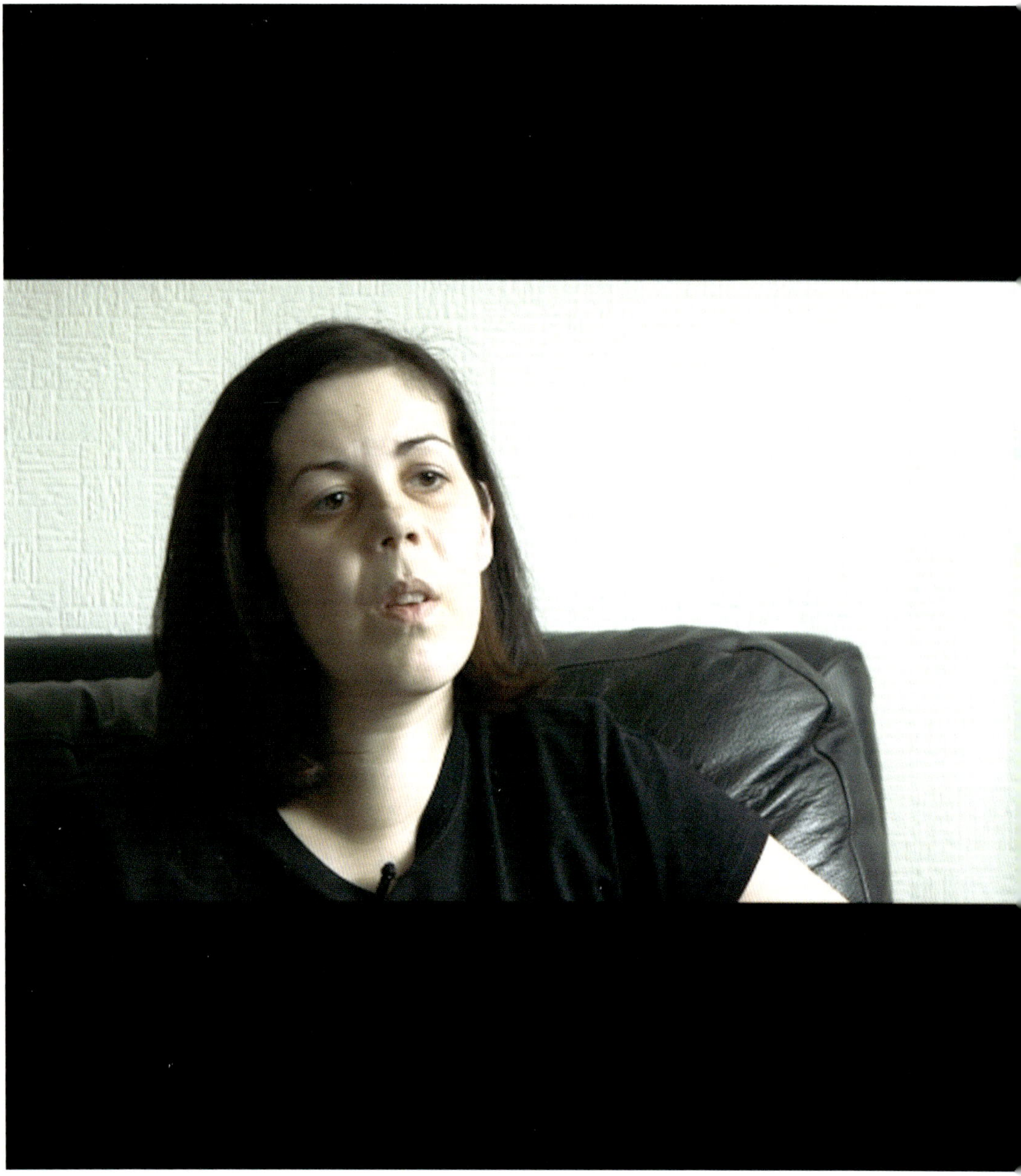

Hazel, 2015

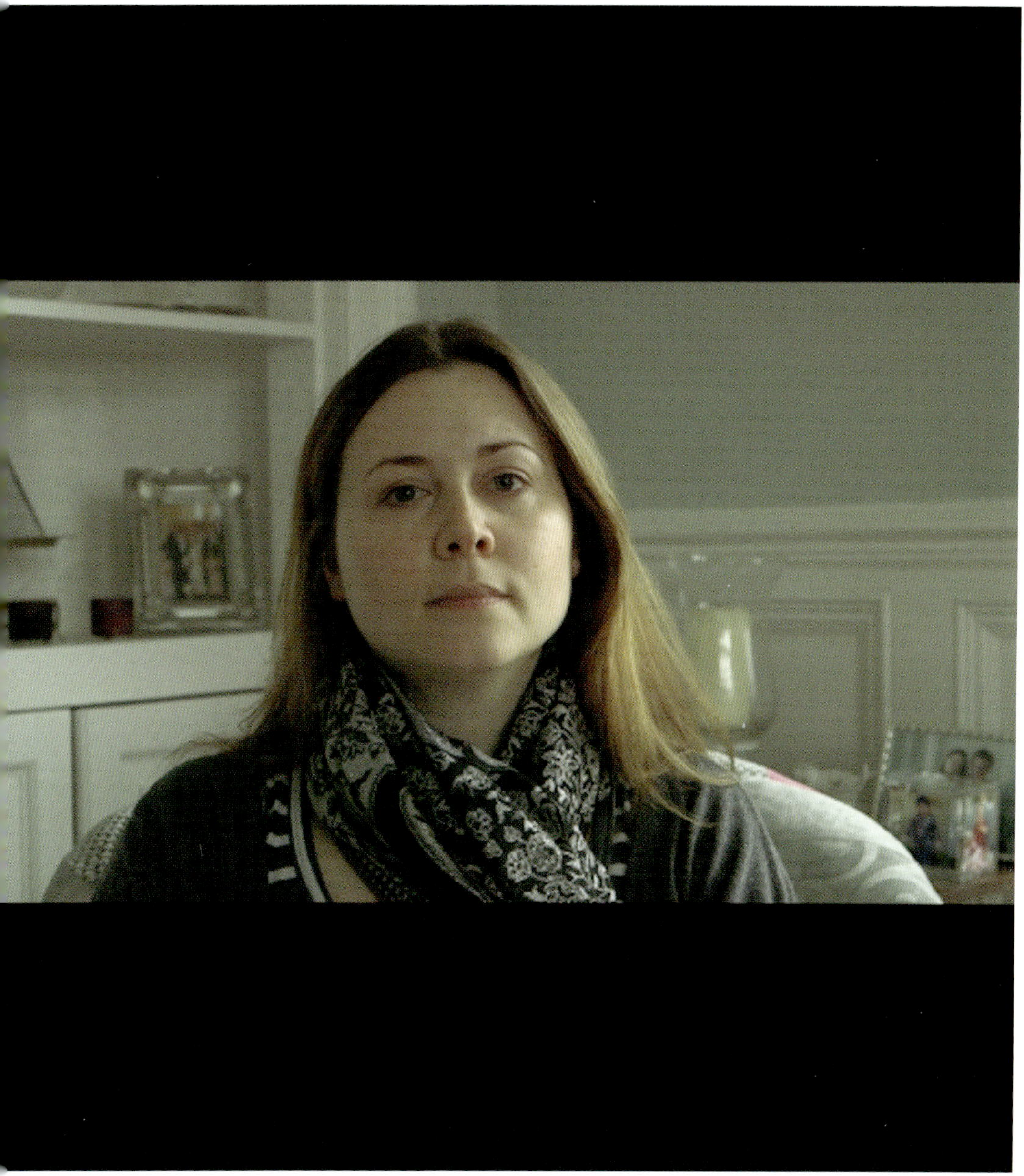

The *Hazel* Interviews

As part of ongoing research, Jacqueline Donachie interviewed sisters whom she contacted through the UK Myotonic Dystrophy Patient Registry. She asked them to describe their siblings, their experience of ageing and their thoughts on genetic inheritance. The women ranged in age from 28 to 68 years old. All of the affected sisters were diagnosed as adults, and in the film the unaffected sister is silent.

The *Hazel* interviews were filmed between 2014 and early 2015, with the support of Northumbria University and the Arts and Humanities Research Council (AHRC).

Credits
Camera: Holger Mohaupt, Jim Rusk
Thanks to: Susan Craig, Libby Wood, Kate V Robertson, Pat Montgomery (MYDY Trust), Chris Dorsett, Mariel Fox and all the sisters who participated.

Louise and Rachel

Louise, 52

Age at diagnosis: 18
Interviewed Northampton, 14 March 2015

Jackie: *Can you tell me your name, and how old you are?*

Louise: My name is Louise, and I am 52.

Can you tell me a wee bit about when and where you were born?

I was born in Amersham in Buckinghamshire. I lived in Beaconsfield for 49 years and then I moved up here to Kettering in 2002, and I've been here ever since. I grew up in Holtspur. I went to Holtspur County Primary School from five to 12, then I went off to Beaconsfield Secondary School. We used to walk to school in Holtspur because it was just up the road and then ride my bike, and then when my mum got a car, we went in the car to Beaconsfield Secondary and my sister went to the same schools as well.

Could you tell me a bit about your family – like, if you have any siblings?

I have a sister whose name is Rachel. She is actually five years younger than me. She doesn't have myotonic dystrophy. Both my parents have died in the last 20 years.

Did you enjoy riding your bike to school?

Yes. I took the cycling proficiency test at 10. The police came in and laid things like firemen's hoses around the school. I remember that. I passed, and I've still got the *Highway Code* that they gave me, signed by them. I've still got it upstairs. The police signed it.

Tell me about the academic side of school. Did you enjoy it? Did you go on to any college or university after?

At school I did CSE [exams] at the time, and I got eight of those. I didn't go on to any colleges or anything. I went straight from school to work. I had a weekend off from Friday [when] I finished school, and on Monday I started work.

What work did you go into?

The first job I had was at a place in Beaconsfield that did analytical instruments – so we had something to do with parts of the Hubble telescope that went into space. I was a clerk for that for 10 years.

I went on to Bateman Chapman which was oil and gas rigs. We were dealing with Piper Alpha – the one that blew up – and I was a typist for that.

After another 10 years I left there and I went to Carless & Adams who were involved in insurance for commercial buildings and also some residential. I was there for 10 years, then I moved up to Kettering where I am now and I worked for a company called AJS & Matchless Owners Club which deals in the spare parts of classic motorbikes, and last year I left there, and I'm now working at another place a little further away in Moulton Park. I'm on reception there. It's called Auto Rescue Logistics and we deal with garages and insurance for getting vehicles off the road after they've been in accidents.

Tell me about your own home life – if you're married and if you have any children.

I've been married to Robert for 22 years this June. We don't have any children, but I do have a dog who is actually banished to the spare room at the moment because he's a bit of a pain, and I go and see friends quite regularly; I'm going out this afternoon and evening as well.

I quite enjoy my weekends. I get my weekends off.

The next few questions are a bit more to do with when you were diagnosed with myotonic dystrophy – but before that, would you be comfortable to talk about whether it was a deliberate decision to not have children?

It just never happened.

I'll ask you a few questions about myotonic dystrophy.

I first realized I had myotonic dystrophy when I was about 18. I found that I couldn't open (I know, it sounds daft) lemonade bottles. I had to use the palm of my hand, and then I couldn't release my grip. I thought, I think I've got this because I knew what my mum was like and she had the same symptoms.

She found out when she was a lot younger, but Mum didn't do anything about it. I was the one who asked for ... I went to the doctor's and asked for checks and Mum was a bit sort of, 'Oh, why have you done this?', but all four of us went to John Radcliffe Hospital and we did a spit test (I remember that) and I think they took a blood test as well. They did check our blood a few years after that and said that it had sort of mutated.

Since I've been here in Kettering, I got in touch with the local support group. I go to their meetings every six months when we have them because we never used to do anything like that. There wasn't ... We didn't know anybody who was down in Buckinghamshire where I was before. I didn't know anything about the actual support group then because Mum just wasn't interested in doing anything about it.

But you said your mum knew she had myotonic dystrophy?

Yup, but she wasn't here. Well, yes, because she found she couldn't get up on her toes – on the balls of her feet. I think she must have been about 39 and she said that when she went to the doctors, and said 'If I'd come to you years earlier when I first realized, would it have made any difference?', and they said, 'No'.

Mum wasn't interested in meeting anybody else. I didn't know there were so many people who had it. I know it's a very common one, but I'd never heard of it before, and nobody's heard of it, and it's a very common disability – but Mum wasn't interested in doing anything, like meeting anybody else, and I'm quite pleased that I have done. I'm glad I sort of put myself forward, and I cope with it.

You find out that you do adapt. Things that I couldn't do, you find ways of doing or if you can't do it, you get somebody else to do it – so, that's the way it is, and I find...

But I am noticing gradually that I can do less and less, but I do actually cope. I keep going as best as I can.

And your sister? Did she get tested at the same time?

Yes, Rachel was tested at the same time but she hasn't got any symptoms and she is actually free from it. Also, my mother had a sister who was unaffected as well, but she didn't have any children. I haven't any children myself and my sister hasn't got any children – so it seems to sort of end with me, really, I suppose.

I'll try and go to the conference, but I don't think I'll be able to make it. It's just sort of time restrictions and actually driving there and getting back here all on the same day – but I have been to one. I went a few years ago with my sister to Cardiff and we found it very, very useful, and I'm going to a meeting in a couple of weeks to meet up with everybody else.

Because you haven't got a wider family who have symptoms, have you found it a useful thing to meet these wider support groups?

Yes, I see how some people are worse than me, and some people that aren't. One person couldn't keep... Her neck had gone and her head would go backwards and forwards. I find that I think, 'Well, I can still hold my head up'. I find that I can't sit up now from a laying-down position, but if you get on your side, you get up – so that's how I adapted. If I can't reach a top shelf, I've got little step stools that I use to get to the top shelf, and if I can't reach, then my husband will do it, or something like that – so, I find that I adapt.

But it is nice to meet the other people anyway just to sort of keep in touch, really. Also, going on the Registry is very helpful because I know if there are any trials going on, I want to be involved – I want to try anything – so, that's why I am on the Registry.

And the fact that the Registry contacted you about our kind of non-scientific thing, like to do this interview?

Yes.

You don't mind things like that?

No, no. It's fine.

How do you feel within your own community about having this condition? You know, you stumble sometimes, and people don't know what it is that makes you do that.

Well, I'm flat-footed. The disability has affected me that I'm actually flat-footed now. Gradually things, I've noticed, have stopped working like my fingertips are very weak, but people say, 'Well, what is wrong with you?', and I say, 'myotonic dystrophy', and they don't understand what that means so you have to explain that the toning actions of the muscles don't work and they gradually will stop working. But when I'm sitting down I look absolutely normal. It's only when I stand up that I find I have to get my balance to get my centre of gravity.

I am very slow with my walking. I don't walk if I can help it. I use the car to go anywhere – but I can walk, but I am very, very slow. But people say, 'Can't you work the muscles?' – like going to a gym or something like that – but it won't make any difference. Once they're gone, they're gone and they won't come back and also, I don't think that going to the gym to work the muscles will make any difference because the muscles will break down quicker than I can probably build them up, and I don't think I'd be able to use any gym equipment with the actual state that I am in with this disability now.

Apart from the physical symptoms, have you noticed any difference in your appearance?

With my appearance (my physical appearance) I know that my face has dropped – especially my eyelids. They are a bit hooded and also, I can sort of pull them and they look a bit strange. My mouth is sloping down, so a smile is sort of straight. I know that I've got a lot of lines as well.

And then my hands – especially my fingers... Sometimes, a lot of people have hands that look sort of a bit like spades, but I'm trying not to do that. I try and keep them... Because for a lot of people they just look like that [*holding hands up with fingers close together and curling upwards slightly*] – like they're sort of set like that. I remember my mum's hands looked a bit like sort of spades, like tilted slightly.

But I know it's definitely my face that's actually dropped. I can tell my features have actually changed. It's all going down. I used to have a chin, and now I'm sort of all neck – if that makes any sense.

And relating that to ageing? Do you think it's all to do with myotonic dystrophy, or are there other elements?

I think with my age – I'm gradually fitting into my face and I think, with my age as well, the disability with the walking is actually fitting with my age. I'm catching up... I think sometimes people will be older... I think, getting older, I'm catching up with how sort of doddery I am – if that makes any sense.

I think it is ageing as well, but I think that it is more advanced because of the physical features – all the actual physical changes – are more advanced because of the myotonic dystrophy. I don't think my eyes would be quite so hooded – I don't think – and, also, probably, my mouth would be slightly turned up a bit more.

Also, if I'm just concentrating, my mouth just drops open. I find I can't breathe through my nose too well. I find that I breathe through my mouth more often. I find that easier.

Do you notice that you've changed anything about your beauty routines that acknowledges that you're getting older? Have you noticed anything different?

I find that I have to keep myself looking younger. I'm using a slightly heavier day cream – but I like to wear make-up every day to go to work because I think I look better with it. It takes ages to look this bad – but I don't wear make-up over the weekends and I do, sometimes, change my hair colour. Being an Avon Lady, I can sort of get the stuff, and I do have a perm. It looks a bit better with a perm. I try and keep myself as good looking as possible.

I don't think I've changed my routine at all, it's just that I wear slightly heavier night cream and day cream, just to sort of try and keep the lines away.

Thinking of that, I'd like to talk to you about your sister. Your sister is younger than you.

My sister's name is Rachel. She is five years younger than me and she lives in Kent. She is about the same height as me, slightly larger than me (she's got big thighs, my sister) – but she's very much like my father whereas I take more after my mother. But we do look a little bit alike. I can see me in my sister when I look at her face. Also, I've actually found that, sometimes, if I've got my face buried when I talk – if I've got like a pillow, and I'm talking into a pillow, my voice sounds just like my mother, and my sister sounds just like my mother as well. And my auntie, my mum's sister, she looks like my mum.

I find that, if I look at people more, I can see my mum in all of us, really – but Rachel is more like my father, but she does have my mum in her as well.

What does your sister do?

My sister works for a private hospital, as a receptionist – so she gets all her ailments sorted out privately.

You're the big sister. Do you see your sister as having aged, now that she's 47? Can you talk about the difference between when you were teenagers and how you are now?

With her being five years younger than me, I didn't want her to be around when I was younger. She always wanted to, 'Oh, Louise told me that; Louise did this; Louise did that...' – but I think we're more getting to the same age now.

I went and saw her for New Year. I went to stay with her for a few days over New Year and it was nice to catch up because we don't see each other from year to year. We just talk on the phone. It was nice to go and see her and to see her changing slightly. She is getting a bit sort of fatter in the face, I think.

We get on fine. Talking on the phone, we can be on the phone for about an hour at a time, and we haven't actually spoken about anything at all – that's the way it goes – but it was nice to go and see her at Christmas and I see her every New Year. She is ageing, but she is not going to be as bad as me.

OK. So thinking about when you were teenagers, I want you to describe your best and your worst fashion moments for both you and your sister.

Oh crikey! Fashion?!

Well, Mum and Dad would never let me wear things that were fashionable. I remember when everybody went to school when I was about 14, 15, in high heels, I was in flat shoes. Mum and Dad wouldn't... And I love shoes! I absolutely adore shoes. I've got loads of pairs of high heels upstairs which I can't wear anymore – but I love shoes.

I think, probably, it was shoulder pads because, being born in 1963, my music and fashion things were sort of the late seventies and the eighties, so I think it was probably shoulder pads in the eighties. I think I've still got something upstairs that's probably still got shoulder pads. I don't seem to get rid of anything.

And I think socks that came up above the knees. I remember wearing those. I think school uniform wasn't really much fun, but I don't think I've ever had any... I did not sort of think, 'Oh, I must go and buy that – that is in fashion'. I've never been sort of very into what people are wearing and I don't take any notice of it now at all. I just wear what feels comfortable and what fits.

Between you and your sister – was there any rivalry or competition when you were growing up?

I don't think we were very competitive. I think it was more or less when I was younger, I think. I remember my mum was saying Rachel would come out with my words – certain things that she said, and she said, 'Where did you hear that?', and it was like, 'Oh, Louise told me'. Now, I don't remember that. I can't remember back to when I was five and Rachel was born. I just can't go back that far. Rachel was just always there. She's always been there.

I don't remember walking home from school one day at lunchtime, just to come home and see Rachel. I don't remember that. My Mum was saying, 'What on earth are you doing here?' I'd walked home – I just walked home from school and didn't tell anybody and came home.

Rachel was a bit of a pain being so much younger than me. It's just – she was there, sort of thing. I suppose, when I started going out with boys she was always in the background and I didn't really want her there then – but it was never like a rivalry. She went out with boys younger than when I started going out – so, we never had [any] sort of bitter rivalry; we never used to fall out – although I did knock her down the stairs once.

She followed me upstairs. I was dragging my space hopper behind me and she followed me up; I did not want her there; I dropped the space hopper and knocked her straight downstairs. I got told off for that.

Any broken bones?

No, she didn't break anything, but she did leave the space hopper in front of the gas fire once and it nearly blew up. Dad found it – luckily. It was Rachel. She wanted to use my space hopper. It ended up looking like it had a toothache because it had a slight bulge in it.

She always wanted to use what I had. Also, I know that both of us – we both had stabilizers on our bikes and we both learned to ride without stabilizers on our friends' bikes. My friend was Gillian and I learned to ride on her bike without stabilizers. Her friend was Nathalie. She didn't have stabilizers and Rachel learned to ride her bike on hers – so, I don't know why both of us learned to ride bikes without stabilizers on somebody else's bike, and then Dad took the stabilizers off.

Would you mind saying a bit about when your parents passed away – how old you were?

My father died the day after his seventieth birthday. He was in my car at the time. Luckily he managed to stop the car and pull over to the side of the road. It was his second heart attack.

Luckily, my mum organized a surprise meal for him at a pub the day before and, even though it sounds strange, both my sister and I had been out a few weeks before and we both bought video cameras. I bought the VHS one and Rachel bought, I think it was Video8 then, and we both did it on the same day, but we hadn't spoken to each other because Rachel was living in Kent at the time and I was down in Slough in Berkshire and we both said, 'Don't tell Mum and Dad' – so we were able to record the party which was the day before my dad died. The day before his birthday we managed to film my dad, so we've still got my dad on film which is really nice. We were so lucky to do that, and we hadn't even spoken about buying video cameras. 'Don't tell Mum and Dad we bought the cameras.' We bought these cameras – Rachel and I.

I got married the year before in 1993 and my father died in 1994. My mum was very disabled at this time. Her glasses used to slip down her nose because her face had completely gone. She couldn't see because she couldn't keep her glasses on. She was at home, but she died three years later – so they'd be gone now about 18 years.

It was a shock. I've lost both my parents when I was in my thirties. I'm talking to people now who are in their fifties and they've still got their parents alive which would be nice. There are things I still want to say to them, and if I do dream about them, which isn't very often, but I sometimes do – if I dream about my mum, there's absolutely nothing wrong with her. I would go back to where we lived when we were children and my mum is there. That's where I'd go; if I do dream, I'm actually in my childhood home and there's absolutely nothing wrong with my mum.

What symptoms were you aware of as your mum got older?

She was slow. She used to trip over a lot. I don't trip. I'm very careful. She did trip over quite a lot. She always had awful trouble with her bowels – I must say that. I'm getting symptoms like that as well. Every time my mum used to go to the toilet, we used to make her a hot-water bottle because she used to have terrible problems and I know that is one of the symptoms. A lot of people think it is IBS, but it is not. It is something else. I cope. My mum was very bad. She did actually go into a home and she did actually die in a home, but she wasn't there for long.

If you have to tick a box in a questionnaire for your age – are you young… How would you describe yourself?

I'm 52. I never lie about my age. I think I'm sort of getting on a bit – especially with the people I'm working with at the moment because the lady I work with is 50, but the rest of the people I'm working with are all in their late teens and early twenties and when I say, 'I've been married 22 years', and they're not even that old, they seem to zone out completely. I feel older, but I honestly don't think that I'm going to live very long. I don't want to get old. As a joke you say, 'I don't want to get old' – but I don't think… As my mum was 67 – I don't think I'm going to get much beyond 70. I feel as though I am, actually, getting older and I don't think I'm going to be able to do everything in a few years' time and I don't honestly think I'm going to get past 70. That is the way I feel because with my mum having had this disability, and me having it, I feel as though I'm going to follow my mum, sort of thing – so, I'll probably get to about 70 and that will be it, whereas my mum's sister who was two years older than my mum is now 82 and still going strong, and still driving – so, it's just the way I see myself, not getting that old.

The thought of ageing – is that a worrying thing for you?

I don't actually worry about getting older. I know it's inevitable. I just don't think I'm going to be able to do everything in a few years' time. I hopefully won't have to use a wheelchair, but I've got a walker in the conservatory. I don't use it at the moment.

My mum used to use a walking stick, but I don't think I need that yet. If I'm out with my friend, we do link arms and off we go. I find that I don't go out very often on my own to walk around – but I can get about without having to rely on other people.

I see other people who are older than me and they seem to be walking fine and I think 'I wish I could do that' – but I find with the disability that I am catching up with my age, or my age is catching up with my disability – so, gradually… If I go a bit slower, people understand that – that I am getting older.

And you don't take any medication or...?

No, I don't take any medication at the moment. And as I say, I don't get headaches or anything, so I don't take any headache pills. I don't take pills at all.

Because your mum already had the symptoms when you were diagnosed, can you talk about your feelings when you found out that you had it too? Had you seen the effects at that point?

I don't think I really noticed that Mum had anything wrong with her. My mum was 32 when she had me and 37 when she had my sister and I think she went to the doctor's when she was about 39 because of the standing-up on her toes and things and she said, 'If I'd come to you five years earlier, would you have known?'... 'So, they never actually sat us down and said what it was, it just sort of was 'there', and I knew the name of it. I didn't know anything about it, really, but then, as I say, Mum told me. I don't think I said to Mum, 'What is wrong with you?' It just was something that Mum had. It was never a 'Sit down, and we'll discuss this' thing.

I just noticed that I couldn't release my grip. I used to thread a needle with the other hand – but I never noticed that Mum was getting any worse, but towards the end... I think when I sort of hit my late teens, then it really did... There was a lot of things going on, and working as well, I was able to speak to people more than at school because it was never mentioned at school, and with Mum doing nothing about it, I wanted to find out what was going on.

Say, when Mum was about 39, she knew about it, but up until she was 50, nothing was done and we didn't know anything about it and Mum was really cross that I'd asked the doctor what was going on; can anything be done? We didn't know if there were any trials going on – so, Mum was a bit upset that I'd done this, but we all went and Dad said, 'Oh – you know – I'm alright'.

But the thing is, when they asked us to do the spit test, if your mouth is completely dry, you can't spit on demand! We were all tested with the blood thing – even Dad, and he had needle-phobia – and he said, 'Oh! I'm not going to be tested', but we went to Oxford, to John Radcliffe. I can remember that quite clearly. I remember sort of snapshots – and I think Dad was quite pleased that we'd actually done something even though he wouldn't say to Mum, 'I think she's done the right thing'. He would sort of like say things.

But then, again, I had my eyes lasered. I had to wear glasses when I was 14 because, I think it might have been cataracts – I don't know. My granddad had cataracts. That was the first time. That is as far back as we can go. When I was 17 and working, I went from glasses to contact lenses, then my eyes went dry and I had my eyes lasered when I was 37 which I don't know whether that... I know that people can have that done. I thought, 'If Prince Philip can have it done, I can have it done', and [singer and TV presenter] Cilla Black as well. But, I've had them done, and now I have to wear glasses to see close up. I don't believe that was a symptom. I think it was just because I'm short-sighted like my mum. It was just one of those things. I don't know whether that was related to anything – but they say that I haven't got any signs – or I might have early signs of cataracts – but I don't know.

Now that I go to these meetings every six months when we hold them, I don't really take much in about all the researchers going off to Paris and wherever they go – it doesn't really interest me. I just want to know if there's anything they can do or any trials or anything. I want to take part and be involved in something.

I can't do a sponsored swim though. There's no way I could do that! They say swimming is very good for you. I haven't been for a long time. The water at Kettering Pool is so darn cold, we don't want to go there anymore, and getting out to another place is very difficult.

I find that I'm very lucky with that – I do have a disabled badge in the car. That goes with me everywhere so I can park where I like. I do get free car tax which is very handy and I really do find the DLA [Disability Living Allowance] money is very, very useful.

Also, I had another banister put in, so I have two handrails to go upstairs, so I'm like that as I'm going up [*illustrating how she pulls herself up the stairs with her arms*]. I have a bench that I can put across the bath and I have a sling round the toilet so I can put my hands down and stand up. Instead of holding onto the hand basin and what-have-you, it's by the toilet, and I can use it to get up with my hands. I also bought a Radar key so I can get into any disabled toilet in the UK.

It's a very long key. [*Laughing*] Every time I go to use it the door is always open! I use the actual disabled toilet at the office. That's downstairs. The other ones are upstairs – so I always use the disabled toilet downstairs at the office.

I don't think you look 52. I think you look really remarkable for a 52. That seems to be quite a characteristic.

The thing is, if you do that... [*lifting eyebrows with her fingers*]

My sister has her eyes pinned up. She's had surgery.

Did she? How old was she then?

She's going to be 50 this year and she's had it done at least five years.

Did it make a difference?

It did. It made a marked difference at the time – but her eyes were worse than yours when she had it done. She found it was making her so sleepy so she got a wee incision at the back and then they pinned it.

She doesn't now look surprised?

No. She doesn't. I think you can also get a thing done with your [*eyebrows*]. *But she had the eyelid thing done. It worked amazingly well at first and then, what happens is it droops down again.*

Is it actually the lid – it's actually this, it's not sort of there? [*Indicating eyelid above fold, near brow*]

They cut a wee bit of skin out of the actual eyelids so if she shuts her eyes she's got a wee scar on each one. So she was really happy with it. She had black eyes for a few days and stitches.

My sister says she was going to get every piece of available surgery to make her look younger until she's 90 – like Botox, plastic surgery – she's always said that. But thinking about all those treatments that are available to you, particularly as an Avon Lady – would you do any of these more drastic things?

I would never do anything like that. I would never have Botox because I think that by the time muscles are coming back, the paralysis is... [*laughing*] Botox paralyses the muscles... I know I could do with getting rid of a bit of this here [*frown wrinkles*] but I wouldn't do that because I think that with myotonic dystrophy, the paralysis is there already, so it won't make any difference. It might take some lines out, but I don't think it would ever come back. I want to be able to do that. I want to be able to at least frown. No, I wouldn't do that.

I've had varicose veins done four times on my legs – but that is probably because of things like that. I wouldn't have anything else done. If I was like that [*sinking back in her chair, holding eyelids open with fingers*] to watch TV in case my eyelids come down, then I probably would have my eyelids done – but I can see fine – now. I think that is the only thing I would do, is probably the eyelids – but that's the only thing.

Myotonic dystrophy has so many symptoms that affect so many parts of you – if you had a magic wand to change one, which one would you change?

If I had a magic wand to change one thing it would be to be able to walk properly again.

Is there anything you want to say about the whole condition, about your thoughts on it?

I don't think about it every day. I've got it; what can I do about it? There's nothing I can do about it – so, get on with it. That's the way I feel. I've never... I don't think, 'Oh, my God ...' It's just one of those things. If I don't have it... I don't really think that people who haven't got it understand.

My sister hasn't got it; my auntie hasn't got it... I know they know what I've got, but I don't really think that anybody who hasn't got it really understands exactly how bad it is – but it doesn't affect me... It affects me in a lot of ways, but as I say, I adapt.

My husband changes lightbulbs that I can't reach. I've got a little step stool in the kitchen that I use to get to the high shelves or something. If I can't do it, I will find a way. As I say, my fingertips have gone; [getting things open] with my fingers I can't do, so I use the palm of my hand. I can't use just one hand. I can start the car engine ok.

Getting cold is the worst part because everything stops working – especially hands. I've always got cold hands, cold feet, and a cold nose – so I try and keep as warm as possible, but you can't wear gloves when you're trying to de-ice the car or turn the engine on. I just try to wear gloves whenever.

I find if I stand up for quite a long time – that gets me and I find that I do... If I'm standing still and chatting or something, I will keep crossing... Also, I cross my legs which I probably shouldn't do and then my legs probably go stiff, so I have to sort of get up, but then getting the balance is a bit... I do feel a bit doddery when I get up.

I like to sort of sit up quite high, if I can. I don't like to be too low because of getting up. Like, if you go somewhere and they've got a low sofa and you can't get out of it!

Also, I went out a little while ago on an old bus and I could not get on the back. I had to grab hold of both the handrail that's there and the pole thing, and somebody had to push me up the bum to get me on that bus. I cannot get on a big wheel. I have to push the thing down to get on a big wheel, and also, they have to... On the fairground – one of those (rides) where they've got four chairs on each thing [*demonstrating a carousel-like movement with her hands*]. They have to put it back to the bar so that I can get in.

If I try to stand up on a chair, I can't get my leg up. I can only get one leg up. I haven't got the strength to pull the other leg up.

I find that I can get in my front door alright. I don't need a ramp for anything like that.

I just get on with it, and that's it. I don't think about it.

Do you think about your summer holidays? Are you going on holiday?

No.

Do you like holidays?

Oh, yes. But I told him I'm not going on holiday on the Norfolk Broads anymore. I can't get off the boat.

Rachel, 47

London, 15 May 2015

Jackie: *Can you tell me your name, and how old you are?*

Rachel: I am Rachel and I am 47.

Could you tell me where you were born and where you grew up?

I was born in Amersham in Buckinghamshire and I grew up in Holtspur, which is a little tiny village just outside of Beaconsfield in Buckinghamshire.

Can you tell me about your family? Do you have any siblings who you grew up with?

I have one sister, Louise, and then there was just my mum and my dad.

And maybe say a bit about your position in the family – who was the oldest; who was the youngest; what kind of age difference there was between you.

My sister Louise is the eldest and I'm the youngest. There are just the two of us and there's about four years' difference in age between us.

Can you tell me a bit about your school days – where you went to school; whether you enjoyed it; and whether you went on to university or college?

For school, it was a local school in [Holtspur]. I really enjoyed that, but when I went to secondary school in Beaconsfield... I wasn't really particularly academic and didn't really enjoy it very much – but from secondary school, when I finished, I went on to college and did, not quite a degree, but a course in business studies which I thoroughly enjoyed.

And then say a little bit about what you do now.

I work now for a private hospital in Beckenham in Kent. I am an admissions officer, so I deal with all the financial parts for all the patients who are coming in for surgery.

OK, so it's dealing with payments and insurance?

Yes, I do. All the insurance companies and all the health insurance companies and if the patients are not insured and they are paying for themselves, I am the one that has to collect the money from them before they can go on and have their surgeries.

That must give you quite an awareness of the cost of healthcare. Perhaps you can talk about that?

Healthcare is terribly expensive and the insurance companies do have to authorize everything at every step of the way. A lot of people, when they take out private health insurance think that, because they are paying for it, that they are going to be automatically covered for everything and, unfortunately, it isn't always the way – so they end up with a little telephone call from myself having a polite little conversation to tell them that they are not covered and that they have to pay and that can then be a bit of a stumbling block because they don't realize that we're talking of thousands of pounds – not a couple of hundred pounds and that can cause all sorts of difficulties. It is generally sort of last minute when I speak to them – so, it's sort of like today for tomorrow and a lot of the patients then struggle to get that amount of money together to continue on the following day.

So how do you feel about that? There's a lot of discussion about the Health Service. A lot of people in this country have no awareness of what their healthcare costs.

No, they don't and the NHS [National Health Service], I think, do a fabulous job under extremely difficult circumstances and the care that they give you is fantastic and I also think that private has its place as well if you are lucky enough to either be able to afford the private health [care] or you can afford to pay for it for yourself, and you can go in that direction, then it leaves the NHS – personally, for me – I think it leaves the NHS to help those who just don't have the ability to pay for it themselves, but they get exactly the same care as they would do if they were in private.

Have you had experience with both?

Yes, I have. I have experienced both.

Can you say something about that?

I've experienced both. In private you have a lovely room to yourself; you have your own bathroom; it's just like your own bedroom – you've got a bed, a television, you have everything you want. The nurses are very attentive. There is still a lot of waiting around and you have to wait for the consultants and the anesthetists. The lists are generally full, so you have your slot on the list and they'll call you down to theatre according to that order.

In the NHS – yes, it is very different. You're not given a room to yourself. When you first go in, you're in a waiting room with other ladies and gentlemen; you're mixing – so you can sit there and you can chat and it's... I'm a very private person, so that... I just felt a little bit out of my depth. I was uneasy at being in the NHS only because I like... I want to be able to shut the door and be on my own and you can't do that in the NHS.

Can you tell me if you're married, and your family situation here – if you have any children or pets?

I'm married. I'm married to Andrew – just the two of us. We don't have any children. We've been married 23 years this year. I love my dog. We have a chocolate labrador and her name is Maggie. She's five. We've always had dogs.

Now I'm going to ask you about myotonic dystrophy and what you know about it.

Myotonic dystrophy I understand, again, because my sister has it and my mum had it. I think it's a horrible disease and it affects all the muscles and the nervous system in the body. It's just horrible.

So, in terms of your sister, could you say a wee bit about when she was diagnosed and how you found out about it, whether or not you've been tested for it?

My sister was diagnosed when she was 21. My mum, my sister, and myself – we went to Oxford High Hospital and had some tests there and my sister's results came back as positive. Unfortunately, on the last visit when we were due to take blood tests, I was poorly and couldn't have my final blood tests – so, it was postponed and I then didn't go on and have the tests until, maybe, a couple of years after Andrew and I got married. I had the blood test then and that came back to confirm that I was not a carrier and I did not have myotonic dystrophy.

In the gap, did you have any thoughts, or did your family have any thoughts about whether or not you did?

My sister always had a feeling that I did not have it because we just don't look alike at all. My sister and my mum were like two peas in a pod. They were just so alike, but my sister and I – we don't look alike at all. I think my sister had an inkling that I didn't have it and I must admit, I probably had a little bit of an inkling as well that it wasn't something that I was carrying, but it wasn't until that final blood test that it was confirmed that I didn't have it at all.

How did you feel then, getting that result, perhaps comparing that to how you feel now, having seen the effects on your sister – the difference?

When the result came through, perhaps just a little relieved. That might sound awful, but it was, 'Ok, I don't have it, and I am not a carrier', and from that initial relief, it was a little bit, 'Oh dear'. I then felt bad because my sister had it and it just reconfirmed that there was this big difference between us – a huge difference

You know, you said it's a horrible thing. Just, if your sense of relief at not having it, has increased, seeing the effects on your sister, or if it's just something that you don't really think about. You probably don't see your sister very often – but when you do see her, maybe to talk about the way you've noticed your sister change because that figure must be quite marked because you don't see your sister every day?

I don't see Louise every day. I last saw her at Christmas and New Year and, yes, we see each other, maybe, once a year – so there is quite a big gap. We speak to each other on the telephone. I have noticed that she is much clearer on the telephone. I think she understands, perhaps, a little better now that she needs to speak slowly and pronounce her words a little clearer.

To me that is no different. I've always been able to understand my sister and my mum. I know a lot of people who perhaps have never come across myotonic dystrophy or don't realize that my mum and my sister were ill, perhaps thought there might have been something else going on and they would always keep asking them to repeat the answers or to say it again – but because I grew up with it, it was just always there. So, as my sister has progressed (and she is getting worse and that is very obvious), when she speaks to me, I don't have a problem with it. My husband sometimes does and other people around us sometimes do, but I don't.

When she was diagnosed, I didn't see any change at all. It's only been in the last sort of 10, 15 years, maybe, that I do notice it a lot more. Whether that is because I don't see her as often or if it is just, maybe she is deteriorating quite... 'Regularly' doesn't seem the right word, but deterioration is progressing.

When we do meet up, when she comes here (she's been here a couple of times), I am very careful and aware of what's going on because, when I don't see her so often, and she arrives, I'm always looking – is the floor uneven; is there something in the way; where are the dog toys; is it safe for her to be walking around; does she need some help; should I move the furniture away? All these things where I've watched my mum go through exactly the same thing and I've seen my mum fall so many times to the point where we have ended up in hospital. You want to take all those obstacles out of the way so they don't hurt themselves. [*Upset*] I'm so sorry.

To know that my sister is going to have to go through everything that we both watched my mum go through; to know that she's going to have to do it, and she's so far away... [*Upset*] Because she knows what's coming, and I think that must be awful to know that this is what's going to happen, and how it's going to be and I don't know if she thinks about that; does she look back on my mum and think that this is what she's going to have to go through as well or does she just blank it out and think there's no point in thinking about it and worrying about it and waiting for it to happen – I don't know.

I feel like coming and giving you a big hug. It's just a natural thing to the conversation. There's another sister I interviewed as well. My sister is the same. My dad had a whole lot of other health issues – yeah, just to know that they have that.

I think, genuinely, there is a bit of you that just blocks that out. I think you see it and worry about it much more...

... than they do.

Yeah. So, thinking about your sister and your mum – you've seen your mum, and your sister must have seen your mum – she said that she goes to the support group meetings where she meets other people and she finds that very helpful. Could you say a bit about whether you've ever met anyone else who has myotonic dystrophy – either in your family or in a wider situation? Have you ever been to any of these meetings, and your thoughts on that?

My sister, Louise and I, we did go to one of the meetings in Cardiff when one of the professors was talking and – yes, it was... I found it a little bit strange to be surrounded by all these people who had exactly the same condition where, previously, we only knew about my mum, and then we only knew about my sister. We didn't come in contact with anyone else who had it. It's not anywhere else in our family – so, theoretically, it will stop with my sister in our family.

It was very... This whole conference centre just full of people who were showing different stages and I remember sitting there and Louise would say, 'Oh, I can see it in her...' or, 'I can see it in him...' or, 'Oh, yes, is that how I sound when I speak?' and, 'Do I walk like that?'

In a way, it's nice to know that you're not on your own. You are not the only person, family, that is going through this. There are other people out there. There are support groups that you can go to. I did go to one with my sister quite a long time ago but because she's in Northampton, it's rather a journey for me to go and see her, and I would quite happily go and support her if she wanted me to. Again, it's nice to know that there are specialists out there who are looking into it and trying to find a reason, a cause. Hopefully, some day in the future, there'll be a cure – fingers crossed.

I know my sister would love to have a cure in her lifetime, but I don't think that is going to happen, unfortunately. I think it is going to take a little bit longer – quite a bit longer to find a cure for this and I do hope that they do.

The research has moved on in the time I've been involved, and I've been involved for 15 years. It's quite remarkable how quickly it's moved on from where they were. So they've got symptomatic treatments which are helpful, like drugs to keep you awake or treatments for your eyes or a whole lot of things. In terms of the bigger picture – it happens because your DNA is changing, it's stretching throughout your life, and that's what makes the symptoms worse, and they are very close now to stopping that happening.

Brilliant – fantastic.

It really is within reach now. There is a kind of endpoint that they think they can get to.

Excellent. Now that I did not know. Again, I will confess, because it's not around me all day, every day and it's put to the back of my mind because it's not in my life, it's in my sister's life, and we're not in each other's pockets, so I perhaps haven't... Because, again, it doesn't affect... Oh, this is going to sound really selfish now, but because it doesn't affect me directly, I perhaps haven't kept up with all the information that has been coming out of the testing and the research that they've been doing. I would imagine that my sister has, but I've not kept up with it – so, that might make me a little bit selfish, doesn't it?

Not at all! I'm going to ask you a few questions that I've asked everyone. I've met your sister, but in a lot of cases, I've met one sister and I haven't met the other one – so I'm going to ask you to describe your sister to me as though I haven't met her.

My sister. OK, she is 5 foot 7, 5 foot 8; she is quite slim(ish). She's got blue eyes. Her face is quite long and thin and she is showing signs of myotonic dystrophy. She does walk rather flat-footed. She does have problems with her hands – opening bottles and anything that might be too tight. I do notice in her face that the eyelids do droop a little.

I find it very difficult to describe my sister. She can be a bit loud at times and I do have to ask her to keep it down a bit, please – but underneath it, she is a lovely... I love my sister and she's... It's a shame that she is living so far away. I'd love to see more of her but it just doesn't happen. Life just takes over. I do find it very difficult to describe her.

Have you seen other people who have the condition? Do you think you look like your sister? Do you see her looking more like people who have myotonic dystrophy?

I look nothing like my sister. If my sister walked in here now and sat down next to me, you would not even think that we were related. We are completely different. I'm slightly shorter than she is; a little bit heavier than she is; different colour hair... We're just so different. The only thing that is alike with us is the fact that we both have blue eyes.

You don't look alike; you don't have any strong family resemblances?

Now that we've grown up from when we were small children, there may have been a few similarities but as we grew up, we just moved to the ends of the poles. We changed so differently – so differently – and I don't think we're alike at all. Sometimes, when I'm talking to my dog or watching TV or laughing, sometimes I can hear my sister in myself – the way she laughs and the way she talks because she has a dog as well, and the way I've heard her talk to her dog and I think, 'Oh! I sound just like my sister.' But other than that, I don't see any resemblance at all. Even my husband has said that he doesn't see any resemblance between my sister and myself.

And did you always feel different when you were growing up? Were you close as children?

I wouldn't have said we were very close as children with the four-year/five-year age gap, she was always that much further advanced than me and I would have said not particularly close – no – not at all.

In relation to ageing, and thinking about how your sister has changed in the last 10 years, particularly in relation to appearance, do you see any difference in your own appearance – if you feel you've changed, now that you are in your forties, or if you feel you're one of these people who just always looks the same – no difference.

I don't think I've aged that much. I don't feel I'm in my forties. I feel the same now as I did when I was in my thirties. I've maybe put on a little bit of weight which, I think, a lot of people do as they get older – but I don't feel any different at all. Not at all.

Great. Could you have a think about your relationship with your sister? Do you think she has aged – not specifically in relation to myotonic dystrophy? One woman I spoke to said that she felt she was older because she did different things now like buying antiques instead of going to IKEA. So, when your sister comes, do you think she seems more mature or have you noticed her ageing?

I don't think I have noticed my sister ageing and I find that she's aged perhaps because of the myotonic dystrophy; because of the effects that it has on her. If she didn't have it, then she wouldn't... To me, if she didn't have myotonic dystrophy, I wouldn't expect her to be, perhaps, stumbling around quite so much as I would do if she was, maybe, in her seventies or her eighties. But I don't see very much change in my sister. She's always been quite bubbly and a bit loud – but haven't we all? I think it's the myotonic dystrophy that is making her come across as ageing. I don't see any ageing other than that.

Your sister is a few years older than you. When you were teenagers, did you share each other's clothes, did you have any style issues, did you think you were cool and she wasn't – anything about that kind of thing – two girls in a house?

Two girls in a house, five years' difference between them as we were teenagers – yeah, I hated my sister! I really did. We argued constantly just over trivial things. Yes, I probably felt I was better than her. We never swapped clothes; I never liked her fashion – but then, being that much taller than me and slimmer than me, her clothes wouldn't have fitted me because I'm quite short. I've always been a little overweight, a bit tubby, so it wouldn't have worked sharing the clothes and the fashion and I didn't think my sister was fashionable at all – just not at all.

It was just the arguing, and my parents told us off time and time again for arguing and we always argued.

Did you have your own bedrooms?

We had our own bedrooms and, no, they were separate bedrooms. We hardly ever went in. I do remember when my sister moved out – the first thing I said to my parents was, 'Can I have Louise's bedroom?', because it was bigger than mine – but in the end I didn't want it. I wanted to stay in my own little room.

She never came into mine; I never... I used to sneak in occasionally, but I never wanted to go in because there was nothing in there that I wanted.

Thinking again about ageing – do you have any thoughts about getting older? Do you worry about it? Is it something that's just happening – you're not aware of it? I have asked this to people in their sixties and clearly that is quite different.

Ageing. I try not to think about it. I do wonder what it's going to be like not having any children or... [*Upset*] And I've got to that point now where I think, 'Will I end up in a home; will I be able to stay here; have I lived more years than I've got left to live?' which I never thought about in my thirties but since I got past 45 I think, 'Oh, have I got another 45 years to go? That's 90. Do I really want to get to 90?'

I do think about Alzheimer's and dementia – is that something that's on the cards for me or my husband, and if it is, how am I going to cope or how is he going to cope? And then I push it away and think, 'Forget it', because, you know, I'm not 50 yet and we'll just see what happens when I get there.

Can you talk about how you think the ageing process is for your sister?

With Louise, it does cross my mind how much longer she's got left, knowing that our mum died when she was 66 and my sister is now in her fifties, and I do wonder if Louise will get to that age and if she doesn't, will she get past it, or will it all happen earlier? And again, she's so far away that I can't help her. Again, seeing it with my mum and how much of a support my dad was, and I want to be able to support her and it's the distance. I know she's only a couple of hours up the road, but our lives are so different and I do worry for her, how she's going to cope and how much longer she's going to be here with us. [*Upset*]

With this platform, is there anything you would want to say to the world of science – to all the doctors, medics – is there anything that you feel is the worst aspect of myotonic dystrophy for your sister or for your mum? If there was a magic wand, that could just take one part of it away, what would it be?

If there was a magic wand and it could remove... It would be nice if that magic wand could just remove it all and just take it away completely, but if that magic wand could only remove a small part, or one part, of the disease, I think it would be how it affects the limbs, the hands, and the feet, and the walking because when you're not walking properly, you stumble, you fall over, you hurt other parts of the body, you break other parts of the body – so it has that knock-on effect. I would go with the hands and the feet if I could have those two on the magic wand – especially the legs and the walking.

Thinking about that, your mum and sister having it and that neither you nor your dad do, did you have ever have any emotions that you associate with how you think about myotonic dystrophy as the unaffected person? For example, some people talk about guilt because they didn't get it, while others talk about feelings of anger (I get quite angry, sometimes).

I am just thankful that I don't have it. [*Upset*]

I do find not being affected by myotonic dystrophy, I am sometimes grateful that I'm not and, again, does that make me selfish that it's happening to someone else and not to me? I don't know how my sister feels. I don't know if she feels thankful as well that it's not happening to me and that it's just happening to her, or does she feel angry that it's only happening to her and why is it not happening to Rachel?

I try not to think about it and, again, does that make me selfish again that I don't because it's not in my everyday life? Sometimes I do feel a bit angry at myself for feeling... I do sometimes feel a little that you shouldn't be like that because it could have been you as well and I don't know how my sister feels that I don't have it.

If you both had it, it would be so much worse.

Do you ever wonder what your sister would be like, and what your relationship would be like, if she didn't have it?

I think if my sister hadn't got the disease, I really don't think our relationship would be any different than it is now. After she was diagnosed and while we were going through our twenties and into our thirties, I wouldn't have said we were terribly close – but I don't think that was because of the disease – it was just how we were. Again, with that sort of four-, five-year age gap she was out doing other things; different friends; she was in work before me; she was driving before me. I don't think our relationship would have been any different than what it is now because of the way we are and the way that we're both made up. It wouldn't have been different.

We talked about clearing things away to protect them (I do), because of the way they walk, and I'm looking at kerbs everywhere because my family do fall and hurt their legs and arms and...

... heads. Yes.

So that relationship of, she is my older sister, but then there is that subtle change where, when we're together, I'm a bit more of a carer, looking after her – have you noticed that?

Yes, when we're together, I do become, perhaps, a little bit more protective. When I'm out and about on my own, if I trip up on a kerb – well, silly idiot – and you just get on with it, but when my sister and I are out together and – yes – you spend your whole time looking down because you need to know where the kerbstones are up or if they're dropped or if there's a hump in the pavement. We do, sometimes, hold on to the arm. I do feel very protective when we're outside and when we're in shopping centres and areas like that where there are so many hazards you can trip up or fall over, bang your head, and there are so many sharp corners on all sorts of items that, if they do trip up and fall over, are they going to bang their head? She just drops. She can't stop herself – so, she just drops, and that's when you think, 'Oh! Has she broken something? Is she bleeding? Are we OK?'

She does still drive which I think is fantastic. I know that's going to come to an end – hopefully not too soon, because it gives her her independence; she is not reliant on somebody else to do things for her. I think she wants to keep going; she wants to be able to have that independence; she wants to be able to get in a car and drive somewhere. If it is just down to the shop for a pint of milk or drive to work, and that she is in work.

With work... I know companies are not supposed to take disabilities into consideration and possibly age into consideration, but with myotonic dystrophy, if you don't understand what's going on, those that are not affected... 'Ok, she's disabled. Well, we don't want her.' But just because she's disabled doesn't mean she can't work. I just wish... I hope more companies out there are going to look at disabled people and think it's the body that is having the problem – it is not the mind and it is not the ability and it is just that independence for my sister, and I just hope she can keep that going for as long as possible.

In terms of your family dynamic now, do you still feel like the younger sister?

I do. I am still the younger sister and – ahh – do I? Obviously, when I talk about my sister, she's the eldest! I wouldn't have said ... I don't think of me as the younger sister. I probably think, now that I've grown up a little bit and got a little bit more mature, I think we're sort of ... I don't see age with my sister as anything. I just think that we're sort 'there' – if that makes sense.

I only talk about my sister as my older sister because she is (to our colleagues and friends) but she's just my sister and I just see her as being sort of there – if that makes sense.

When you are together and a family decision has to be made, who makes them? Do you work together, or do you defer to your older sister when you have to make joint family decisions?

Joint family decisions... We haven't had to make joint family decisions for a very long time. I'm thinking back now to what we did. I think we would try and be together. It was never a case of 'Louise is the eldest, therefore she takes the lead in most things.'

When my mum was alive, not long after my dad died, even though my sister was living with my mum at the time, my mum always looked to me for the financial assistance – so, I was the second signature on the bank accounts and I made sure that (even from this distance) the bills were paid, but my sister was living in the house. I don't know if that's because my job is financial with the insurance companies and the packages in the hospital, or whether my mum felt that I was more sensible and could deal with all of that even though we were living sort of an hour, hour and a half, away from each other. I don't know if my sister ever felt, 'Well, shouldn't it be me because I'm the eldest and I'm living here?'; but my mum was more comfortable with me having that responsibility. I don't know if my sister was, 'Yeah, that's lovely! Thank you very much – I don't have to deal with it because someone else is dealing with it.'

But, when it came to, obviously, the funerals for my mum and my dad, we pulled together. It wasn't a case of, 'Well, you're the older sister – you do it'; or, 'You're the younger sister – you've been looking after it all for so long, you do it'. I think we came together as a unit and dealt with things together and we sort divided things up equally amongst us rather than me taking the lead or me saying, 'Well, you do it'. We would look at what needed to be done, and we sit there and say, 'Well, I'll do this and this – what about you do that and that?' That's how it worked. I think we came together as a unit for those family decisions.

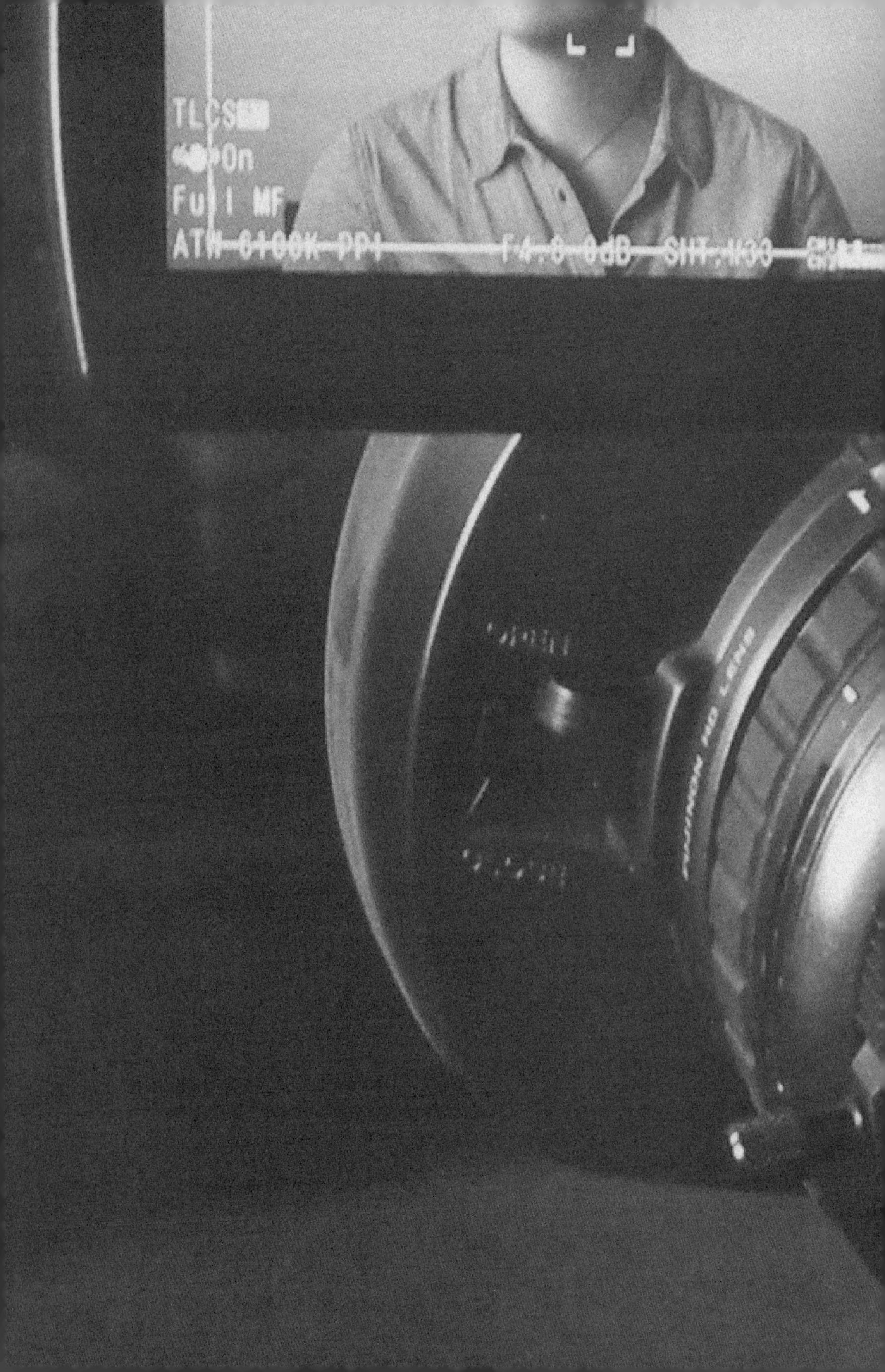
TLCS
On
Full MF

Jennie and Pam

Jennie, 68

Age at diagnosis: early 40s
Leicester, 16 March 2015

Jennie: My name is Jennie. I'm 68 years old.

Jackie: *Can you tell me about where and when you were born?*

I was born in Lincolnshire in a market town called Grantham, which is the birthplace of [former UK Prime Minister] Margaret Thatcher (for those who are interested). I was born after the war on 2 December 1946.

Do you have any siblings?

I come from quite a large family with gaps between. I have, at the moment, three living siblings, but initially there was my brother who is nine years older than me. He is still alive and very well. There was a brother between my older brother and me. He died in infancy, then there's me, then there was a brother, Trevor, who was four and a half years younger than me. He died in his early thirties. Then there's my sister who is eight years younger than me and she is living in Australia and doing very well, and is very healthy, and then I had a younger brother who was one of twins. The twin died at birth. He died in his early forties. So, quite a large family, really. We started with seven, and today, altogether and surviving there are three. Quite a big difference. Only one of my siblings didn't inherit the myotonic dystrophy gene.

I would like you to tell me about your childhood – where you went to school; if you stayed in school; what you felt about those years of your life.

I was born and grew up in Grantham. I stayed there until I was in my early twenties. I was 21 when I left. I remember it as a very sleepy market town, very boring. Even at four and five I think I thought it was very boring. It was a small place then, and obviously much bigger now.

I went to the local infants school as it was in those days. It was called [ ] Infant School. It was in quite an old, austere building and was made of stone. It was post-war, so everything was very scarce – not a lot of things, and certainly school life, for me, was very different than it is for my grandchildren today.

I went on to a junior school and then on to secondary school and left school before taking my O levels [exams] (as they were in those days) because my mother had given birth to twins (one died) and because we were a large family, there was really a lot of pressure on me to help and that wasn't conducive to studying.

Maybe it was a mistake on my part, looking back, but at that time I thought it was the right thing to do – just to give up; probably find a job; and help because I'd have more free time to help my family. A job was nine to five, and studying was, obviously, weekends and evenings, and I just couldn't keep up with that.

Looking back now, I had a very good memory and I should really have stuck it out. So I went back later on to do O levels, A levels, and got a BA Honours in Applied Social Studies from Trent University in Nottingham and, combined with that, was a social work qualification. I wanted to do social work having seen what had happened, by that time, with my family. I felt that I wanted to be able to support and work with people who had some difficulties. At that time however, I didn't know that I myself had difficulties.

Let's talk about your relationship with your sister and when you were both still living at home. How did she view you?

My sister doesn't have the myotonic dystrophy gene – so, therefore, her children don't have it which, obviously, for her, is a great relief. She has expressed some feelings of guilt because she feels that she has escaped the inheritance pattern. So, she feels relief, I think, but also feelings of guilt especially with so many people in the family dying over related issues to myotonic dystrophy. I think she worries about me and my brother, and everyone in the family who has this condition.

Growing up, we were eight years apart – so it makes a big difference. She was a plump eight-year-old and I was a very skinny 16-year-old – so, at that point, I think she would have liked to wear my clothes, but there weren't many that she was able to wear because it was the time I was in short, tight skirts and long boots and different things. There were some things I'm sure she borrowed.

She loved to go through my drawers and wardrobe and she loved to play around with make-up – but I think it was when we were quite a bit older that we swapped clothes and different things. Maybe she was borrowing my clothes, but she wouldn't have been allowed out in them. My mother wouldn't have allowed that. So, yes, it was the time of miniskirts and high boots and all of the things that have sort of come back into fashion. I think being very slim, and being able to wear those clothes, I think was very good.

When I was 18, my sister was going through puberty and was a bit chunky. Maybe for her she would have liked to borrow my clothes but she wasn't able to get into them. Not then, but there was a period when we both were the same – when we were, obviously, quite a bit older.

I think I would buy her things when I was at work rather than that she would borrow my things. I worked in retail at that time – so, I would end up buying something that I thought would look fine for her. She thought that was really cool.

We've always been very close. Maybe because we're two girls in the family, I don't know – but also because we're very alike even though we don't look alike, our personalities – what we love to do, things that we like are so similar. Even though she lives on the other side of the world, we have a very strong connection.

So thinking about how far away your sister lives, and considering that you only see her every year or every couple of years, do you notice differences when you do see her that you would attribute to ageing?

Obviously, I didn't used to see my sister a lot. I used to try to go every second year to Australia. When she was first out there, I didn't see her for a long time. She came back a couple of times when my father was ill and dying and then when my mum... So we didn't see a lot of her. I didn't go over at those times, but once I was out working and had my own money, I was independent and I could go and visit her.

We used to send each other photographs, which we don't do anymore because now we have Skype. Now we probably see each other around twice a month – so she hasn't really altered very much.

She's quite dark – rather like me, going a bit grey. She still has quite a round face because she doesn't have this condition. If you saw a photograph of her when she was, say, 12, 15, 20 – you would still recognize her today. She hasn't changed that much.

How would you describe your sister?

At one point we looked quite alike, but not anymore because she doesn't have myotonic dystrophy – so she's kept a round face and all the things that you have when you don't have this condition.

My sister and I, although we physically don't look so alike anymore, we are very alike in the things we like, her sense of humour, what programmes we like, what books we like to read, what music we like, the sort of clothes we like, where we like to go on holiday – we're so similar, although we are eight years apart – we just now look different. So I can buy things and send them to her, and likewise, she can do the same and we know that we're going to love whatever we get – that's very fortunate.

She's a nurse in Australia. She's been nursing since she was 17½ and did her training. She's lived in Australia since she was in her late teens, early twenties. She loves the life out there. She'd been married three times. From her third marriage she has two children. Her son is out there and doing very well. Her daughter is over here teaching and doing very well. She doesn't have myotonic dystrophy, so her children cannot have myotonic dystrophy.

She does a lot with her life. She tries to get over to England as much as she can. Obviously, she wants us to visit her, but that is not so easy anymore.

I think a few years ago we had this plan that when she retires (which won't be for quite a while yet), that we would spend time... I would spend our winters in Australia, and she would spend Australian winters in England so we'd have summer all the time – but that maybe isn't going to be quite like that anymore. But she still comes over to visit.

She loves going to the theatre – so, when she's over here, we try to do all the things that, maybe, living in a small rural town in Australia, is harder to access.

Let's talk about fashion. I've asked everyone else – what are your feelings on shoes and have you had to change which shoes you can wear? For example, heels?

One of the changes, fashion-wise, is not so much with clothing, but it is with shoes because I've got a heart condition, I've got a pacemaker, and I've got something called cardiomyopathy. One of the things is, you get your ankles and feet swelling, and when it's hot or humid it's even worse. That has been a big change. I always had very slim feet; I could wear anything – and now I can't. I have to wear shoes that are adjustable so that when you have fat days, they will allow you to wear them with comfort, and when you have thin days, you can tighten them so that they stay on your feet. So shoe shopping has become a big problem.

I don't have drop-foot yet which you can have with this condition – but I do have to be careful of not stumbling. That happened earlier this year and it was a bit of a disaster. I don't want to break anything else, so I'm very mindful now of finding shoes that fit properly – so, no heels, not too flat (because that causes problems as well – it needs a bit of an incline), sensible, and comfortable.

If you're thinking about how you would classify yourself in an age category, like young, middle-aged – how would you describe yourself?

When I think about having to define yourself age-wise today, I would think that can be quite difficult for everybody – not just me. Usually, there's a nice little box where it says, '65+'. That's the one I tick, I'm 68.

However, that changes if you're talking to a 10 year old. My grandchildren think I'm very ancient. They ask me the most amazing things, did I go on a stagecoach, or was I in the Civil War – or whatever! But if you're talking to somebody who's in their forties or fifties, they don't see themselves as that far away from you – so it's a moveable feast – ageing. I put myself in the '65+', heading on to 70, and God willing, I will get to 70.

What are your thoughts on the future? You're aiming to get to 70, how do you see the next 10 years, perhaps?

I think I just take it a day at a time. Certainly, over the last 20 years, my health has gone downhill with a big wham because something's happened. I've either been ill (nothing to do with myotonic dystrophy, I had something else), and I don't bounce back very well. So, that, and obviously this latest episode in my life with having heart problems, has been a real eye-opener. So, I take life a day at a time and I'm grateful that I feel I'm back on the up again, thanks to acupuncture, and thanks to the person I go to see.

I'd like to be active and healthy – as healthy as I can be...

Let's talk about your working life.

During my working life, I worked as a social worker in child protection which was long hours and I really feel for people today because, way back when I was doing it, there was still a shortage of manpower and overloading of cases, and quite dire situations that you were trying to help families work themselves out of, and in some cases, they weren't able to do that – so there was always some heart searching about the next step and what would happen.

Then, I guess, the last 12 years of my working life I decided that I would like to get into work that would kick in earlier than the child protection bit – so I went to work for Child Mental Health Services. That, I think, was where I did find the work that I really felt was right for me.

Families were coming because they had themselves volunteered to come, not because they were made to come, and they worked very hard to get themselves out of what we all can get into – quite difficult situations.

I worked with families in family therapy; working individually with children; and I feel quite proud that I think a lot of families who might have ended up in the child protection arena didn't because they found the inner strength within themselves to resolve the issues that they were struggling with, and I like to think that I did help.

If you could talk now about your own family. You have two children, do either of them have myotonic dystrophy, and if so, when were they diagnosed?

I have two children. My son is now 45, and my daughter is 43. My son lives in Australia with his second wife. He has two daughters, one by his first wife who lives over there, and one by his second wife. He doesn't have myotonic dystrophy, but he was in the Gulf War and has severe Post-Traumatic Stress Disorder and latterly has been diagnosed with bipolar disorder because of the problems that went on and on. It's very sad, he's changed from being a very outgoing personality to being somebody who is, I think, quite afraid to go out – so that's been quite catastrophic.

My daughter has myotonic dystrophy, but has a wider gene expansion than myself. She lives in Leicester. Part of the reason for me moving here was to give her some support because she has two young children and also so that, if things got worse for me, that I would have somebody too.

So, it went from me coming to support her and the children to her having to give me a lot more support. Her mobility is quite poor now. She uses a walking stick all the time; she has just been awarded PIP [Personal Independence Payment]; she has two children – 10 and six – and she helps her husband in the management part of his business which is in electrical work, so she is a busy lady – obviously gets very tired and pushes herself very hard, tries to be a good mum, tries to support her husband, and run a house and all the things that women have to do.

I help out with the grandchildren – not as much as I was doing because I've had problems with my health recently, but hopefully I'm back on the upward move and hope to give her more help in the future.

Can you talk a bit about when you were diagnosed with myotonic dystrophy?

When I was diagnosed with myotonic dystrophy, we were already aware of it in the family. Initially we weren't. My brother, who was four and a half years younger than me, had joined the RAF and began to have problems with his hands. He couldn't pull the trigger on the rifle, so he went into different duties – he went into office work, then couldn't use the typewriter. He was then discharged as medically unfit. After quite a while he was diagnosed with myotonic dystrophy but, as a family, we had no evidence that anybody else had this condition at that time. It was suggested it was some sort of a rogue gene.

My father had, what we now know were, some mild symptoms – so, although he died of cancer, until he developed cancer, he was a healthy strong man. My brother, however, his health declined and he died in his early thirties.

My youngest brother was then diagnosed because he clearly had some learning difficulty and some physical difficulties – not as pronounced as his older brother. When he was diagnosed, I went to my GP to say, 'Look, this is not some rogue gene. I've now got another brother'. I was sent for some tests but [was] told I was fine. So, at that time I would be in my late thirties – but I did give a blood sample. I was then called back a couple of years later because they had made a breakthrough with chromosome 19 to be told that I did have myotonic dystrophy.

My daughter, who came with me for support, was asked whether she would like to be tested and she was just 18. She agreed, and when we got the results for her, she was just devastated.

I think I had long suspected that I did have myotonic dystrophy. It was still a blow to be told that, but it wasn't such a blow because I had already suspected that I did have it. I think the devastating thing for me was that I passed it on to my daughter – so getting that news at almost the same time was the worst bit by far.

Can you say a bit about your retirement?

I retired three years ago on health grounds. I had decided that I was 65 and I would be working part-time. I knew I had this condition but, at that point, I was finding I could cope – but I knew I needed to go part-time but the social workers who worked in Child Mental Health (in Lincolnshire, at least) were actually made to go back into the child protection arena and, unfortunately for me, I spent a whole day working on IT skills and, because I have the condition I have, something went radically wrong. I travelled all the way from Nottingham to Lincoln; spent all day operating, using a mouse, and when it came time to drive home, I couldn't move my arm. By the next morning I was in agony and, subsequently, after I was signed off work because I couldn't do anything – but quite some time later, I was diagnosed with a bilateral frozen shoulder – but in the meantime, because I couldn't drive and couldn't get to work, I've had to give up work, so I never did get back into child protection.

You've recently moved house.

I lived in Nottingham from 1982 until last year and loved it there. I loved everything in Nottingham. It was my sort of place. My daughter and her family live here in Leicester, and I've got a son who lives in Australia – so it was a toss-up: do I go to Australia, or do I come to Leicester, because my health had begun to decline by then. I think it just fell to being Leicester – so I moved here a year ago. So I've been here exactly one year.

I stayed with my daughter for the first... I moved in January, but lived with her for a few months because I had a pacemaker inserted a few days after moving here, so I couldn't actually move in and get on with this place for a few weeks – so I moved in sort of eight, nine weeks later.

A lot of work was done on the house. My son-in-law, bless him, put all his electricity skills to good use, but also his skills as a kitchen and bathroom fitter and made it into a very nice house for me.

With treatments moving on and more research happening, is there one specific thing that you wish you could change about myotonic dystrophy – one symptom?

I think if there was one thing I'd like to change about the way myotonic dystrophy affects me, is in some way to find a way of strengthening muscles because, for me, it's the overall weakness that is creating all the problems for me – whether that would be the same for everybody, I have no idea. But just not being able to do things – open things, safety things for children – I can't open them either. I know that applies to a lot of people, but it's just the general weakness that actually creates some of the problems because that increases your likelihood of falls, not being able to manage your personal life, not being able to do things, like things you need to do in the house – keeping the house OK, or cleaning your windows or whatever – because you get tired so quickly because the muscles are not firing as they should. So, for me, that would be the one thing I'd like to see come into being. If there's a magic treatment out there that will stop the muscles from becoming weak.

Would that be the thing that you would change for your daughter as well?

It would definitely be the same because her mobility is now quite poor. Her mobility is more affected than mine, for her, some way of either stopping her muscles becoming weaker, or actually strengthening it, would make such a difference to her life. She has two young children and she's just in her early forties, so all the things that you would want to be doing with your children – running after them, playing games – she can no longer do.

She is not in a wheelchair yet and for that, I know, she's very grateful – but there's just so much that she can't do. So, it would be on the top of my wish list – for everybody, I guess, not just for her and me.

Is there anything else you would like to add?

I think the message I would like to give to the general medical people we will come into contact with – not so much in research, but for the people I come into contact with who either don't know very much about myotonic dystrophy, or have just not come across it before – I think all of us appreciate just how busy they are, but this is quite a complex multi-system condition and my feeling is, I think, not because people aren't caring, but just through lack of time, they don't have time to ask pertinent questions and lots of things don't get picked up (certainly from my own experience) because relevant questions haven't been asked. It's only when all of these symptoms become a huge problem that they and you become aware of exactly what we're looking at now, and you go away and you research it, and you can quite plainly see that there is a progression of this and all those checklists that would have indicated that this is what the problem is, are there, but (a) they don't have time to ask, and you don't give them that information because you don't know they want it because you don't know what it is that's happening.

But also for the medical profession to be aware that we do know what's happening in our bodies. We might not know why.

So we were at the point where we were talking about what I would like to happen.

The other thing I would like to happen is that, again it comes down to the medics having enough time – but to listen to what we're saying to what's happening in our bodies. We know what's happening. We can describe. We might not know why it's happening or what bit is causing that, but we do know. You become very sensitive to what is happening within your own system and all too often that is dismissed.

You say, 'This is affecting me in that way', and you're told, 'No, that can't possibly be the case'. 'Well, you're not in my body, and you don't have myotonic dystrophy, and you don't know'.

Pam (South Australia), 60

Interviewed Leicester, 24 August 2015

Jackie: *I'll just ask you to say your name, and how old you are.*

Pam: My name is Pam. I am almost 61. I live in Renmark in South Australia, and I work at a small country hospital as a registered nurse.

Can you tell me a bit about where and when you were born?

I was born in Grantham in Lincolnshire in 1954. I was born in a hospital. I was supposed to have been born at home, but I was breech, so I got taken to hospital partway.

And you grew up in Grantham?

I grew up in Grantham. We lived in a terraced house until I was seven years old, and then we moved from there. My mum (I didn't know at that time) was pregnant with twins, so we moved from the house that we lived in to a house with more bedrooms. I didn't know she was pregnant. I didn't know she'd had a baby until I saw her because she went into hospital because she had pre-eclampsia. My sister left school and was looking after us because my dad obviously had to continue working.

So, yes, I grew up in Grantham which, back then, was a sort of semi-industrial, semi-farmland type-place.

I went to high school; I got five O levels [exams]. I actually wanted to do design, but that did not happen, so I went to be a nurse. I went to study nursing and when I was 21 I went to Australia shortly after I'd registered as a nurse.

That's quite a big step. You were quite young when you went. Can you say a wee bit about that?

I went out with a fellow that I'd met and fallen in love with. His brother was already working in Whyalla which is a town in South Australia – so, he was out there working; his mother and father went out, and it was sort of a progression that we went out. So we were all living in Whyalla. I worked as a registered nurse there.

Unfortunately, the relationship didn't continue. He decided he wanted to go off with another lady, so I had to decide what I was going to do. I did not want to go back to work in England! Back then, the salary was really quite abysmal. You could barely make ends meet – so I stayed where I stayed and I don't regret living in Australia.

So how long have you lived in Australia?

I got to Australia November 1976; I came back briefly in 1982 for three years, basically because my father had died. I came home to look after him when he got ill. He had a thing called oat cell cancer which is quite an aggressive form of cancer, so I came home for about three months to look after him and got him home from hospital. Jennie, obviously, was helping me to look after him. I had to go back, so he died probably a couple of months after I returned.

Then I thought I really wanted to be a midwife and, at that time, I was working as a charge nurse of an orthopaedic and surgical floor and the director of nursing didn't want to release me from that position – so, this lady was holding up my career trajectory, so I applied to be a midwife in England and I got into Leicester and Nottingham, and I can't think where else – but I chose Nottingham because it was where my sister lived, and it was about 30 miles from Grantham.

So, between '82 and '85 I studied and worked as a midwife and I had to make a decision then. I had to decide, do I go back to Australia because, at that stage, I was an Australian resident and you couldn't then be out of the country for more than three years as you'd lose your residency. So, I was thinking, 'Shall I go back to Australia? I don't want to stay here'. At that stage I was a pretty smart cookie and I wanted to be a midwifery tutor. I thought, 'Hmmm – pay is bad; I need to have a car; I need to have somewhere to live; I need to be somewhere where I can make money'.

I thought about going over to the Emirates but, unfortunately... I applied. I think Esso had a compound where they wanted nursing personnel but at that time I think an English girl fell out of her window or was pushed out of a window and that frightened the life out of my mother – so I decided not to do that and back I went to Australia and not long after that I met up with this fellow that I subsequently married and had a couple of babies with. So that was that. That's life in a nutshell.

Are you still a midwife?

I no longer work as a midwife, the reason being I work in a very, very small country hospital. It's a minimum-staff hospital where we have two staff on a day shift and an afternoon shift, and a night shift. There are no resident medical officers, so whatever comes through the door, we have to react and respond to appropriately.

I think about seven years ago, it was decided that they were going to move all of the midwifery services to one hospital. I live in the Riverland which comprises of five small towns dotted along the river – so, Berri, which is the next town – that's become the major hospital, so obstetrics was moved there; we no longer have a 24-hour A&E service (it only happens between 8 in the morning and 6 at night)...

I didn't want to go and work in a different hospital, so I decided not to... I'm a registered mental health nurse as well, so I do use some of that in my daily practice. I'm reasonably happy with what I do at this stage in my life. I don't want to do any more training.

Can we just talk about your siblings? What is your position in the family?

I was born in 1954. My younger brother, Andrew, was born seven years later. I didn't know at that stage he was one of twins. I think that would have been the second twin, and that twin died. I don't know whether he had myotonic dystrophy, I don't know whether they were identical or non-identical twins.

So, there's Andrew – that was the youngest one, then myself. My brother, Trevor, was three years older than me. He had myotonic dystrophy. My sister, Jennie, is nine years older than me. Then there was my brother, Tony, who was much, much older than me. I think he was about 16 years older than me. After Tony was born my mother was pregnant and had a little boy called Michael. I don't know if he was a cot death or what happened – but he died when he was a few months old. That knowledge is given to me from older family members. I have no recollection of who he was, really. That's my position in the family. I'm the last-but-one of Tony, Michael, Jennie, Trevor, Pamela and Andrew – so, five siblings, two of which were deceased at or around birth.

Can you describe your sister to me?

My sister is called Jennie, she is 69 years old. She worked as a social worker with a specific background in child protection and child mental health. She is divorced. She has two children, a son and a daughter who are in their forties. Her son lives in Western Australia, her daughter lives in Leicestershire.

Jennie had to leave school when she was 16 because my mother was pregnant with, at that stage she thought, only one baby. She had to go into hospital because she had pre-eclampsia, so Jennifer's education was aborted at a fairly early age, and once she got married and her children were growing up (they were at primary school), she wanted to avail herself of an education.

She had worked in retail and doing various jobs that she could to help augment the coffers – the household bills – and then she did her O levels and her A levels and then went on to do tertiary education. She did a degree in social work, and was a very successful and senior clinician until she retired about two years ago.

I think she's an extremely strong woman. I think she is a role model. She is somebody to whom I aspire. She is kind, she is considerate. She is very observant of people; she uses the skills that she's learnt as a social worker and as a child protection officer, often to point out things to people. I can spot a paedophile now quite easily simply because she's assisted in my education. So, for me, that's useful, both when I was a mum of younger children, but also in my professional life.

What else can I tell you about her? She's extremely tidy. Both she and I grew up in a chaotic household. I think we both feel that Mum was quite post-natally depressed, overwhelmed, and found it very hard to keep law and order in the place – so we ended up trying to keep tidy our own little bit of space. I think it's very important when you are surrounded by chaos to have your own bit of space that you can keep tidy. So, as a consequence of that we both... I don't know that you could say that we've got OCD, but we're probably a bit military in our tidiness. I can't function in a mess, and she's worse than me, I think. She certainly can't function in a mess.

So, for me, my sister is somebody who is a person I can rely on; I can confide in; whose opinion I value. She is also a person who has myotonic dystrophy, and she is a person who has kept herself extremely fit to try and manage and offset any of the complications that the illness is going to put upon her and it really hasn't been until the last two or three years that you can notice that it's having more and more of a deleterious effect upon her since she developed a heart-block and had a pacemaker put in, she's not as fit. She can't do as much. She's a lot slower and she finds that extremely frustrating.

We just had a holiday in Jersey. We had to make sure that we stayed in a place that didn't have lots of stairs; that had a lift; that the terrain was gentle, because her daughter came along whose myotonic dystrophy is a lot worse. So there were so many things that we had to factor into our stay that other people would not think about to make sure people were safe.

Yes, she's finding it frustrating, I think, and she's also... I think she's also sad. I think she's sad that her children have myotonic dystrophy. So, I can see that – and, then, of course, I don't have that. I cannot have any true insight into having a congenital condition because I don't have a congenital condition – so, you do feel like an observer of... No matter how much you can, as a nurse, identify with what people's needs are when they have got mobility issues, it's not lived experience.

How many years has it been since you last saw Jennie in person? Have you seen differences in her over the last 10 years?

I last saw my sister a little over one year ago. It's now September 2015 – so I saw her August 2014. We had gone on holiday to Ilfracombe. At that stage she hadn't had a pacemaker put in. It was quite hilly there and she found it quite difficult to walk up and down the cobbled lanes and alleys – so that sort of curtailed our enjoyment.

Physically I don't feel that there'd been a significant change until you watch her try and do fine manipulations with her hands – opening and closing things; getting something out of a packet – those things I observed. I could see that it is more of a struggle for her. Getting up and down stairs – more of a difficulty for her.

Looking at her, I didn't see any major change. She didn't look fatter or thinner; her face... Her face – possibly muscle wastage around her mouth looked a little worse, and I could see that, sometimes when she was tired, she couldn't articulate as well. Her speech wasn't as clear. It was a little bit slurred at times. I also noticed that eating seemed to be a little bit more difficult for her. She wasn't chewing. Her mouth didn't seem to move as well as I remembered it doing.

So, yes, I did notice a difference then. And we do see each other on Skype – thank goodness for Skype. I think, possibly, had I not seen her on Skype (because I'd seen her two years prior to 2014), I think I may have been more shocked. I may have noted more changes because, of course, when you don't live with a person, they become frozen in time. My nieces and nephews – I can find it so easy to remember them as 12 and 14 – not 45. And the same thing happens for your siblings. They become frozen and you select those memories out because they're the easiest ones; they've got greater longevity. You've lived with them; you've grown up with them – those changes are really strange when you come back after 10 years and see something different.

What was life like growing up with your sister?

Some of my earliest memories in the little terraced house that we lived in until I was seven years old, I have an enduring memory of going upstairs to bed... Maybe I should talk about the house. It was a traditional two-up and two-down – so, in our family, we had Mum and Dad, Trevor, Jennie, and myself. Tony was away in the armed forces, Andrew hadn't yet been born. So we all slept in one bedroom. There were two double beds in there. In the other bedroom was my grandpa (my mum's dad) and Uncle John – so we had quite an extended family – almost Italian.

I can remember being taken upstairs to bed and my sister would read to me *The Water-Babies* and I just loved that story and when I fell asleep, obviously, I was then moved from Mum and Dad's bed into the bed I shared with my brother and sister and I can still smell the damp wallpaper because I was next to the wall and this wallpaper with these faded roses and just damp plaster. That's an enduring memory – probably the most enduring one because you forget so many things.

I can remember her playing with me in the garden. I had this thing about walking around with my mum's old handbag and I would put these big, hairy caterpillars in there, and I can still remember doing it.

So, she was older. She played with us. We had a play shed that was full of old furniture – so we would pull out this furniture – chairs, tables – and we'd make stagecoaches – Jennie, Trevor and myself. So, we had the best times. We didn't have Game Boys. We had to make our own fun.

Then, moving on to being older, she got married when she 21 – so, by that time I would be 12 – so, developmentally, that's a chasm. When you're 12 and when you're 21 – a big chasm. So I was still a little girl when she got married.

We moved to another house so, again, it was still an extended family. My grandpa died not too long after we moved. He died from pneumonia – so, Uncle John had his bedroom. In the front bedroom was a double bed with Mum and Dad and there was a single bed with me in it. There was a cot with Andrew in it, and then up in the attic was my brother Trevor – so, pretty well packed in.

When my sister left home when she got married, I got her bedroom. Oh, I'd died and gone to heaven. To have your own space because, remember, we grew up in a chaotic household. You'd open the drawers and it was like jack-in-the-box – you couldn't find anything. So, sad me, I would spend my Saturday afternoons when I was 15, 16, 17, taking out my jumpers, folding them up and putting them back in again because I just loved being tidy! Isn't that sad?

Anyway, yes, I left home when I was 18.

When Jennie got married at 21... Again, she was something I aspired to because Jennie was slim, good-looking, little feet, little hands – and I wasn't. I didn't get the genes for that. I would look at her and think, 'Oh, she is so pretty! She looks nice in those trendy clothes', because, at that stage, [model and actress]Twiggy was here so thin was good. Thin was good. Mother didn't like thin. She wanted us all to be quite fat because, if you're thin, as a mother, you're not feeding your kid – so she liked these chunky kids. So she was something I aspired to. I wanted to be like her. Genetically it was not going to happen.

I can remember going to stay with her at her house when she and Barry were married. She was a fantastic home-maker. She had brilliant flare; the colours were good; it was clean and tidy and organized. So, I've got very many happy memories of looking up to this person as my sister, as a role model, as somebody I wanted to be like.

And, more so, because she was denied the opportunity of an education – so, for her to then go back once she'd had her children and done that early mothering stuff, for her to be able to go back and carve a profession for herself in times where it was difficult for people and specifically difficult for her because I don't think that her husband was particularly supportive of her having an education. I think he felt a bit threatened (but that's only my opinion).

So, yes, big role model and continues to be so today.

I think, because she has a condition that I can see would affect her a lot more if she didn't look after herself; if she didn't say, 'I've got to stop doing this'; or, 'I need to eat healthily'; 'I need to get some exercise'; 'I need to have some physio' – if she sat and just let it happen to her, she'd be a whole heap worse and I don't think she'll ever do that. So, that's my sister. I love her to bits.

With Jennie as the older sister, how have your roles changed since childhood in terms of looking after each other?

I try, because I'm a nurse, very hard not to take over things. If I see her struggle with something, I will offer, 'Can I do that for you?', but I wouldn't ever presume to take something out of her hands and do it because I think that's insulting – so, if she needs help, I'll offer it. If I can see that she needs help, I'll say, 'Can I give you a hand with that?'

Obviously, there are things we do while I'm here – like, I've hung pictures today. We bought some garden statue things and we carried them. We got plants. She said, 'While you're here, these are the things I want to get done' – so, we do them to the best of my ability. I'll do the heavy lifting; we're going to be painting the shed; I've hung pictures; I've done gardening; I've got shrubs to pull out. I'm delighted to be able to do those things because when you have left, when you've gone, you've left tangible evidence of something you've done.

We were talking about that today – my sister and I, and I said, 'Do you remember when Tony came to visit me?' – my brother, about six years ago now and the plug on my sewing machine (not the bit that goes into the wall, the bit that goes into the machine) had got a loose wire and he fixed it for me so, for many years, I've pulled out my machine, and it just makes me think of Tony. So, I hope that that would be the same for her. She would be able to look at a picture, plus, I made the picture. So she's got double reasons that she'll look at that and say, 'That's Pam. Pam did that for me'.

It's good to help each other. What goes around comes around.

So you never get that tension about being bossy? You're very good at managing that?

I think it's disrespectful. I will only offer help if she looks like she needs it or if she said, 'I can't do this'. I wouldn't... I think it's very rude to just steamroll somebody because they've got a condition. If she's not managing, I'll help her.

I think we know and love each other enough to be able to be honest with each other and if I grab something that she was trying to do, I think she'll tell me to butt out. I don't think she sees me as bossy and I don't think I see her as bossy. We're just good together.

Can you talk about how you feel age has affected you with your physical appearance?

In 1968 I had an accident. I fell over at school going over a vaulting box and, unbeknown to me, had a slipped [capital] femoral epiphysis and that just means that the gristly part on the end of any long bone is where we grow – where that long bone grows from – and I landed badly and I knocked it off the end of my femur where it attaches into my hip joint. So, I walked around on that for a while and my leg stopped growing and, eventually, I kept taking myself to the doctor at 14 saying, 'I've got a pain in my knee and a pain in my hip', and he really wasn't very interested.

It wasn't until a teacher from school summoned my mother to the school to say, 'What's going on with your daughter? She can hardly walk', because I was walking a mile to and from school every day, that she took me to see a doctor and, I guess, having an adult with you validated that, and I then spent 12 months in hospital in a big pair of plaster trousers. They manipulated the head of the femur back onto the shaft of the femur and that's what they did in 1968. Now you'd have a pin in it and you'd be up in a fortnight and you'd get on with your life – but I didn't.

So, really, because I've had a hip injury, that's made life a bit difficult for me. It's not a congenital condition – I can't give it to anybody. It's mine, but it means there are things that I can't do. My femur doesn't articulate properly in either hip because of the way it was treated. Back in '68 that's just how they did things. This made life very difficult for me. I can't ride a horse.

Now, as I'm ageing, I have osteopenia which is like the precursor of osteoporosis so I'm at risk of getting a fracture. So that not quite petrifies me, but it certainly puts me on notice of what I'm doing because I have to look after myself – so I don't want to put myself in harm's way and fall over or fall off something and fracture.

So, for me, the ageing process, in one respect I think, 'Wow, I've got from being 14 to almost 61 without needing to have a hip replacement'. It's worn OK. It's painful and I limp, but it's worn OK, and then I think, 'How much longer do I think it's going to be OK, and what else can I do to be careful to keep myself safe?'

From a muscular-skeletal point of view, that's my little bag that I carry around with me.

As far as other ageing processes go, yeah, I think you notice that your collagen is going a bit and your skin isn't as elastic as it was, you're getting your mother's chin, you got grey hair, you got cellulite, you've got varicose veins – and you think, 'Well yeah – so what? I have and I don't care'. I don't look at it! Nobody else looks at it. The dog loves me anyway.

I think, also, as you get older, you're more aware of the wear and tear. Not just the physical, 'Am I looking gorgeous, still?', but the physicalness of ageing – the fact that you don't have quite as much stamina. Cognitively you start thinking, 'Oh, I've forgotten that person's name!', or, 'I can't remember what I need to do next' because, certainly at work, sometimes you're rushing and it's very busy – there's only two of you; there's nobody else to help – so you've got a lot on your plate and you think, 'I've got to do this, this, and this...' Sometimes you stop and think, 'Ah! I've just walked out of that patient's room. Where was I going?' And you think, 'Was I like that 20 years ago, or is this because you're cognitively less capable?' To me that's a little bit of a worry as a nurse. And we have to work until we're 66 before we can retire in Australia – so that's an issue. It is an issue for me.

What else about ageing?

I think I'm quite comfortable with it. I don't feel like I want to rush off and have some Botox or things lifted and tucked or whatever people want to do. This is my body. It's carrying me around. It's working OK. I don't want to enhance it in any way. Other people can do that.

That probably encapsulates what I think about ageing.

Are you happy with your body?

I'm happy with my body. It does the job. We would all like to look like [actress] Bo Derek when we're 50, but the truth of the matter is, you've got what you've got; you've got your genetic predisposition to be this size. I'm OK with my body.

I'm much more interested in what's between my ears and what comes out of my mouth than whether somebody thinks I'm a size 10 or a size 15.

Can you talk about what you wore growing up?

Clothing-wise, Jennifer had a nice figure. She was very small. She was probably about size 10, and back then the fashions were very slim, very short... She had good legs, so she always looked gorgeous in everything and she'd got a good eye and still continues to have a good eye. She can pick things out that you will look good in.

We have just done some shopping recently with me coming over and she's very helpful looking in the sales to find things.

For me, fashion-wise... Well, I was a bit of a chunky monkey. I spent 12 months lying down in hospital – so, what can you do? You eat, don't you? Your mother brings Mars bars in – so you get to be quite fat and then you've got to get rid of all that.

So, for me, growing up in the seventies, that was my time to try and get some of the blubber off. I didn't really think much about clothes then. It really wasn't until, maybe, 1973 when I started nursing that I took any notice of it.

We used to wear shoes with platform heels. I have got some photographs of me in quite short skirts and I'm thinking, 'Oh, you didn't look too bad back then'.

Maybe because you're a nurse and you're wearing a uniform for the majority of the time, I don't know that I spent a lot of time thinking about clothes. I couldn't afford to buy them anyway, there wasn't any money. You got paid for a month; they took out your board and lodging and food and you got very little left to play with – so, you know, what do you do? Do you go down the pub, or do you buy clothes? Oh, I'll go down the pub, thanks.

I can see why you like Australia!

I admit it, I admit it. I'm keeping the red wine industry going.

Did you ever kind of borrow each other's clothes? Were there any incidents of rivalry?

Hardly! I don't think there was rivalry... She was older and I think she, in many respects, became our caregiver. She looked after us. I couldn't fit into her clothes anyway. Pipe dream. It wasn't a consideration. I don't think...

I mean, I liked the clothes that she wore and I would have loved to have worn them...

I'll tell you a funny story. When I first came out of hospital, Jennie was pregnant and she made me a dress and she tried it on to see if she thought it would fit me. It fitted her, and she was about to give birth – and it didn't fit me! It was a bit tight. So that was, 'Oh, my gosh. I need to do something about this'. So that was a catalyst for me. That was the only time she and I ever wore the same clothes!

Let's go back a bit and talk about your family and when they were diagnosed. When you first found out about myotonic dystrophy, what did you understand it to be, and how did it come to be diagnosed in your family?

Ok. So, thinking about diagnosis of myotonic dystrophy in my family – I had gone back to visit my family. I can't remember the year. Maybe I was 25, and I saw that my brother, Trevor, was walking in a peculiar way. He was sort staggering with his pelvis forward a little bit, trying to stay balanced – almost like a heavily pregnant woman would walk because the centre of gravity changes – and it bothered me because he just looked like he'd been on the beer, and I couldn't work out what was going wrong. So I spoke to Mum and she didn't think there was anything wrong. I really can't remember whether I spoke to my sister about it. I know that when I returned home, I started going through my books and I came to the conclusion he had some sort of neuro-muscular disorder. I didn't know anything about myotonic dystrophy.

I then found out he'd been diagnosed. At that stage we were told that he was the only person. It was a rogue gene because my father was well and fit, my mother was well and fit, nobody else manifested any symptoms, so they said 'Oh, he's got a rogue gene'.

Then, when I came home a number of years later and saw Andrew with similar problems, that was a bit scary. I can remember going with him to the local GP and he had some degree of intellectual incapacity, but he was also quite socially gauche. He couldn't speak up for himself – so I thought it would be good if I went because I have a medical background. He was happy for me to go. We didn't really rehearse what he might say when he went in. Maybe we should have done that because I can remember him sitting in this office with his GP saying, 'I want a full medical check-up', and so this GP was not impressed and told him he couldn't possibly be coming here without a proper appointment for a full medical check-up, so I then tried to explain to him that we had myotonic dystrophy in our family – my brother had it, and this brother is also showing some symptoms. We need to get checked – and that happened. He was then checked, and he was diagnosed. So we had two people in the family with myotonic dystrophy.

I think sometimes you don't join the dots. Sometimes you don't think, 'They've got it; has anybody else got it?' I think Jennie got checked first. I was pregnant with Lauren, I think, who is now 26, maybe 25½. So Jennie had her test and some of them took longer because they were happening over here and the National Health Service has more people, it is a bit more burdened, and some of the tests I had were quicker even though I started later on.

We got through to almost the last lot of testing and I think it might have been the muscle biopsy and she had thought she hadn't got it, and then she found out that she had – so I'm not finishing off. I'm still going through some tests and I'm pregnant with my child and I'm thinking, 'Oh...', and then, when I found out I didn't have it, it was, 'Oh, what a relief', but 'Oh, my God. That means I'm the only person!' And that does your head in. It does your head in because you don't quite know what to think. You're almost ashamed of thinking, 'Phew! I've escaped'. But you're also feeling really sad that everybody else hasn't, and they've got children, and they've passed it on – so, yeah. That was a difficult time.

Do you think your family were surprised that you didn't have it, or perhaps with Jennie, that she did? Was there any sense of, 'Of course, you don't have it...'?

I think, fairly early on... And, again, I wasn't there, and phone calls are very expensive, to place a call from Australia to England – very difficult, very, very expensive – so, I didn't have those big chats with Mum. I know from what Jennie told me that Mum was saying, 'Oh no, it's not me. There's nothing wrong with my family'. Dad was dead by then, so we couldn't do any tests for him. So it was a process of elimination. She didn't have it, so it must be Dad because we are assuming that our dad was our dad for all of us children and there weren't any other dads involved along the way!

I think she was very relieved that she wasn't the person that had caused this. But because you're away, you're not there for great blocks of your life, it's almost a conceptual thing. It's happening to somebody else a long way away; it's not my lived experience and that makes you feel more guilty because you're not there helping; you're not there seeing and doing.

For me, that was difficult, and continues to be difficult. You're not there; you're not seeing it; you're not living it; you're not supporting.

And when you come back, do you feel that difference magnified as, perhaps, the symptoms are increasing within your family?

I do. I've said to people who know me about my family history. I say, 'Really, I'm never going to whine about anything because I'm lucky. I may have a dodgy hip, but that can be fixed and I'm not going to give my kids a dodgy hip'. But for my sister and my brother who have children with myotonic dystrophy, and you can see what the trajectory is going to be like. I feel so sad for them, that they have got this poisoned chalice to deal with. To me that's sad, and knowing that it's going to get worse until – until what? You know, what is going to happen to this little girl; what is going to happen to this little boy; how are they going to be? That, to me, is horrible because, as a mum, you want your kids to be OK. You don't want them to have... You want them to just have acne and boyfriend trouble. You don't want them to have to deal with a lifelong disability that's going to shorten their life and kill them.

Do you ever wonder what your sister would be like if she didn't have myotonic dystrophy? Or if the kids didn't have it? What if?

When I think about my sister – would she be different if she didn't have myotonic dystrophy? She really didn't have huge symptoms until, perhaps, the last three years – so, I never thought of her as a person with myotonic dystrophy. I knew she had it, but it was so apparent that you had to filter that through your interactions with the person. She was pretty able-bodied; we did things together when I was here – no problem. She was as strong and fit as I was.

Whether she would be different? Who knows? She may... When bad things happen in your life, often you will become a better person. You get the opportunity to shine – so she may, or she may not have – I don't know. She's just my sister and I think she's pretty clever and the fact that she now has something that's shutting her off, that makes me sad. That makes me sad for her because, as we get older, there's less choices for us. You think, 'I'm getting older; I don't have a husband; what will happen when I need looking after; I've got these children who have this too – how much longer can I help them for?' These are conversations that we have – whereas, that's not something I have to consider.

I don't have to think about – apart from the normal ageing – when I'm 80, am I going to be in a nursing home? If so, please make sure I've got clean nighties. I don't have to think 'My child won't be able to give me clean nighties because they're going to be in a wheelchair'.

Yes, I don't have those considerations.

You've got your own siblings. It's quite a big family. Is there a wider family – are there cousins?

Well, no – again, this is second-hand information for me.

I know that my father was illegitimate. I'm not sure who his father was. I know that his mother had him. I don't know whether she had another child that died – possibly so. She then got together with another guy and had a whole bunch of other kids. Dad didn't have anything to do with them because they were not nice people.

So he wanted to protect his children. We had to stay in and play with our brothers and sisters. We weren't allowed out and so we had these very hothouse, sort of cloistered, formative years. 'No, you're not going out to play because if you're going out to play, there's kiddie-fiddlers out there and that's not going to happen to you, kiddo'.

I didn't know any of that. I just thought 'I've got the meanest parents in Grantham because we couldn't go out to play! I don't want to play with my brother anymore! He gives me the shits. Take him away!' But we couldn't. We had to stay there.

So, I never met cousins. My mother – she had Uncle John. Uncle John lived with us forever. When she was growing up, her mother left when, I think she was about 12 (again, this is not my history – this is given to me, this information). I think that her mum left because her father was a bit fisty. She went off to work as a housekeeper. I think she then married this fellow but didn't let on that she had my mother and my uncle as children.

My grandpa took up with another lady and they lived together – so, Mum was left as, I think, about a 15-year-old living in her dad's house with her younger brother who was probably about five. So she reared him. He was her little boy. I think she was quite a vulnerable young woman. It was war-time. She was a very pretty lady – a very pretty lady. Subsequently she met Dad; got pregnant... Back then it was like, 'Oh, my God! Pregnant before you're married!' So they got married.

So, I don't know any of her family and we were not allowed to know any of Dad's family – so, I don't have any. It's all like, 'Oh, what a surprise. Big black hole there'.

Who knows...? And we don't think, I was talking to Jennie about this. We don't think Dad's mum was the person that gave him the myotonic gene. We think it might have been his father because when she then went on to have many other children, none of them were affected in any way. So we just deduced that. That was that.

Thinking again about Jennie, if you could change one symptom, which would you prioritize over others?

I think now the cardiac conduction system because it's affected her heart. She has a pacemaker in – so, if I was going to change anything about myotonic dystrophy, from a personal perspective (not for everybody), I would have liked for her not to have had a problem with her cardiac conduction pathway so she didn't need to have a pacemaker in. Because that's been the most difficult thing for her – having that pacemaker in and then falling and whacking her nose and her chest, and then she got so crook [ill] after that and she can't fathom why. She thought she'd fractured her sternum; she thought perhaps the pacemaker had been knocked off-kilter. They said they tested it electronically and it was OK – but ever since then she's not felt as good. They put her on some medication which nearly killed her because she had the symptoms that meant she needed to have a beta-blocker and an ACE inhibitor – classes of medication that treat conduction difficulties in the heart – but they were not good for her. They made her worse and it really wasn't until she saw a cardiac specialist who specialized in myotonia, he said, 'Don't ever take those. Stop taking them' and she had already stopped because she knows her body better than any specialist.

So, yes, for me – I would like to get rid of that conduction issue for her. I think life would be a lot easier then.

And looking at your niece and her children, do you see it as a generational thing – that the things that affect one generation are quite different?

Looking at Gail, she is large. She's got the muscular hypertrophy. That makes it very difficult for her. She doesn't walk very well; she has quite an unusual gait; she's walking with a stick now; she's eating – the way she moves her mouth is difficult. You can see that eating is a bit of an effort. She's got young children; she's tired; she finds it hard to get up and down steps.

My brother who has myotonic dystrophy – he hardly has any symptoms that I can see, yet his daughter has quite marked symptoms. She gets very tired. She can't grasp things.

My brother's son, Michael... I haven't seen him for many years. I hope to see him this visit. He used to live in Australia. When I last saw him – probably about six years ago, I could see in his face that he's got myotonic dystrophy. He didn't talk about it. Jennie tells me now that when I see him I'll be really shocked because his face looks quite drawn; he looks quite unwell. Not easy.

As you age, can you imagine getting any treatments for ageing?

No! What a crock! I cannot imagine, as I get older, having anything that would enhance my beauty because I'm a firm believer that your beauty shines from within and you can be the prettiest boy or girl on the block, and you can be the biggest, hollowest pain in the butt. If you're forever looking in that mirror and making sure your skirt's straight and your hair is nice, you don't get chance to really have a personality – so, no. I cannot imagine.

I'd love to have a hip replacement that didn't give me any pain and was never going to dislocate, then have that leg the right length so I didn't have scoliosis and walk a bit funny – but that's not cosmetic, that's mechanical. So that would be the only thing I'd like. I don't think I need anything else.

I said it to my sister, 'You know Jennie, if I ever won the lottery I would have just an ordinary car with KMA on the number plate which would be "Kiss My Arse"'. I'm not interested in looking gorgeous.

Is there anything you want to say?

I just really hope that with the research that's being done, that they will be able to find out how they can fix this. To prevent it, yes, rather than having to terminate pregnancies, but also to fix it for people who have it. Can we unglue things; can we make it so people can move better; can we make it so their muscles don't continue to hypotrophy. I don't think that's going to happen in my lifetime. But to be able to fix things would be brilliant.

Everybody wants to fix Huntington's chorea; they want to fix multiple sclerosis – it's not necessarily going to happen.

But myotonic dystrophy – scientists love it. There's a lot of movement. I agree it's not imminent, but they are starting to do something, but that doesn't help what we're doing today.

I think being the only sibling in a big family that's got big evidence of myotonic dystrophy, you do think that you have got a responsibility as a relatively healthy human being... I think, when I say 'conduct yourself appropriately'... I think as a nurse I think, 'I have to conduct myself appropriately'. I would never hold the profession in disrepute. I'm a bit of an old fart there. You're not going to see me rolling out the barrel in a public place – but I think as a person who doesn't have a condition that's so awful, I've got a responsibility to make the most of my life and try and help people make the most of theirs – to effect a positive change.

So, when I pop my clogs, I am hoping that I would be remembered for good things, not for mean things, because I'm a lucky woman. It's just a toss of the sperm, really, when that all falls down, which chromosome are you going to get.

Thanks very much.

You're very welcome. It's been a pleasure and a privilege to have a conversation with somebody because nobody knows what it's like, when you feel guilty that you haven't got it, when you feel relieved that you can't have given it to your children, when you're whining about, 'My kid's done this...' and you think 'Oh, God! What am I saying! Her child is never going to do that!' And that, to me... Don't whine.

It's not our fault that we were unaffected. We didn't say, 'Excuse me! Pick me!' It just happened. You get something else instead. You might get bowel cancer or ovarian cancer – none of us get out alive. We're all going to get something. So, hmm. I'd like to die of boredom, really.

Michelle, Stephanie
and Lorraine

Michelle, 36

Age at diagnosis: 20
Essex, 17 March 2015

Jackie: *If you could tell me your name and how old you are.*

Michelle: My name is Michelle and I'm 36 years old.

Can you tell me where and when you were born?

I was born on 28 November 1978 in Basildon Hospital in Basildon, Essex.

And where do you live now?

I live in Basildon, still, with my husband and my two children.

Can you tell me a wee bit about when you grew up – whether you had any siblings and a wee bit about your family life growing up?

I grew up in Basildon with my mum and dad, and I have two sisters who are quite older than me. My two sisters are Stephanie, and she is 44 now – so she is quite a bit older than me, and my other sister, Lorraine, she is 40... coming up for 42, so she's a bit older – so I didn't see them much. I didn't get to play with them as much because they were out with their friends while I was quite little.

Can you tell me about your school – what kind of school; if you liked it; and when you left school and if you went on to any other college or university or got a job?

I went to an infant school and a junior school that were combined – so it was like, go to the infants, and then onto the juniors. I liked school. I liked going to school, and I liked every sort of subject, really, and then I went to secondary school which was just basically next door and I liked most of my subjects. I wasn't particularly good at a lot of subjects, but I liked it, except my last year at school. I wasn't very happy with my last year at school. Although I went and did my exams and everything like that and then I stayed on for a year in sixth form and did an NVQ [National Vocational Qualification] in health and social care, and then I went on to a training sort of college, I suppose, where I did childcare and education.

I had a placement in a school where I was like a trainee teacher's assistant and I absolutely loved that. So, after I'd done that course and passed and everything, I looked for jobs in schools. I couldn't get one, so I ended up going to a hospital. I worked in a hospital with adults which I liked. I was there for nine years – but I did like that.

Can you tell me a bit about your own family?

I'm married to Simon. He is 43 – almost 44, and we've been married for 12 years. We live in a house with our 15-year-old son and a seven-year-old daughter.

We've got two fish. We did have a dog, but because of Hannah's transplant, we weren't allowed to keep her and my mum has her – but we still pay for her.

We're quite happy in our house. Hannah has a bedroom downstairs because she can't... Well, she can manage the stairs, but she's not as mobile as my son.

Tell me a wee bit more about your children and this ties in with the question, what do you know about myotonic dystrophy? Maybe tell us a wee bit about how you found out about myotonic dystrophy, and what you know about it.

Tell us how you found out about myotonic dystrophy in your family.

I had a daughter at 20... 2003 and she was born early at 32 weeks. She was born floppy and they transferred her from our local hospital where she was born to a London hospital where they diagnosed that she had myotonic dystrophy. At the same time they clinically diagnosed me with the handgrip tests and the reflexes-type thing. I later went for genetic tests – but that was later on. They said they'd diagnose me at the same time which was quite difficult. That was November 2003. She died in 2004 – January – and so we knew about myotonic dystrophy.

So my husband (as clever as he is) decided that he would look everything up that he can find on the Internet. We spoke with a genetic counsellor and she explained everything she could, but a lot of it is really difficult to take in – all that, at the same time, especially when I had to tell my sisters and my mum and dad and they were all tested as well. Well, one of my sisters was. So that's how I found out that I had it. It was 2003.

Then, going on from there, it wasn't really... It didn't really seem to affect me that much because we'd lost our daughter. It was just 'get on with life' sort of thing, and then in 2008 I had Hannah. We knew that she was going to be born with myotonic dystrophy because I had previous tests at 10, 12 weeks. The test failed four times, but then it came back and said that she would have it, at which point I was not willing to terminate the pregnancy at that time. So she was born. She was born at 34 weeks and was on the ventilator for 51 days – but she pulled through.

Just to say what you know about myotonic dystrophy – how you get it, is it inherited and tell us how you understand that you got it, and how your daughter got it.

I understand that myotonic dystrophy is a genetic condition, that it is hereditary and it's a 50/50 chance whether you get it or not. I got it through my dad. My grandparents had died before we knew about the condition, so we're not sure if it came from my dad's mum or my dad's dad – but we suspect it was my dad's mum because she complained about being ill a lot of the time and my dad's dad died when my dad was 18 – so he was fairly young. My nan was like 90 or something

– but she did complain of pains and aches. We actually joked that she was a hypochondriac but now, looking back, you can see what we think it is. They don't count repeats[1] over a thousand in the UK – so, we know I've got over a thousand repeats, but we don't know how many and because my dad's older (because it gets worse as each generation passes it on), my dad only had cataracts and he just recently got diabetes which is part of the myotonic dystrophy thing. I'm pretty sure he has ECGs yearly as well. He doesn't complain of pains or anything like that, or tiredness or anything like that because he's 69. He gets tired anyway.

But he doesn't complain about anything or anything like that, but my sister – my older sister, I think she was in denial a lot of the time to begin with, that she had the condition. But now she's retired from work, it seems to have got worse really, really quickly. She uses a stick all the time. She lives in a normal house and they had to put a handrail up because she's got drop-foot and she falls over quite a lot. When she's upstairs, she'll get everything she needs for the day and take it downstairs so that she doesn't have to go upstairs again for the rest of the day until bedtime – although she does sometimes go upstairs if she needs to for any reason (because she's got a downstairs toilet as well).

So she tires a lot, but she pushes through the pain which I do as well if I'm walking. I get very painful legs. Because I never knew about it before, I just thought my legs were aching because I've done a lot of walking or where I was working, walking up and down the ward. I just thought that I'd done a lot, but they said it is actually the myotonic dystrophy causing the myotonia in the legs.

You can say a wee bit about your sister – I know she didn't find out when she had her daughter, but maybe say a wee bit about that.

Stephanie had a daughter at 29 weeks. She'd been in hospital a few days before and because she'd had a bleed, and they said that she wasn't in labour. They said, 'You can go home if you want', and obviously, you'll go home. So she went home but then rang me that night and said that her waters had broken, so I went round and tried to encourage her to get in the car but she wouldn't, so we, in the end, called an ambulance and she had the baby at home – breech. The baby wasn't breathing, so I breathed into the baby until the ambulance came and they sort of wrapped her up and used the thing to get stuff out of her mouth but there wasn't anything because I'd already cleared it and they took her away and then my sister had to deliver the placenta, so she did that, and then they took her off to the hospital and I drove and picked my mum up because she was asking for my mum and we went up to the hospital and were there for quite a long while. Her baby was in special baby care, but only lived for 19 hours because she was so weak and so early.

Because she was so premature, they thought she'd died because she was premature and breech. They said there was lack of oxygen to the brain and so they turned the machines off after 19 hours and she died, but they never did perform the autopsy

1. Author's note: the gene causes sections of DNA to literally repeat themselves. The number of 'repeats' someone has tends to indicate worse symptoms. In the US everyone knows their 'repeat length', but in the UK people aren't told.

or any tests because they didn't think they needed to and then, when I had my daughter at 32 weeks, and then when she died, we did the autopsy and tests and they took samples of some of her organs to test. I think there are some papers with the information about her on them. Obviously, there's no names, so you'd never know who it is, but they did a lot of research and so, when I'd had Hannah in 2008, six months later another baby was born at the hospital with the same condition and then six months after that another baby was born and because of everything they'd done with Holly and Hannah, these babies did really well.

One of them was very, very severely affected and that baby would have died if they hadn't had all the information and the doctors had been made to read a lot of literature up about it so that they were on the ball for the following babies. That's about it.

Could you tell me a bit about Hannah?

Hannah was born early at 34 weeks but, because of my previous pregnancies, I managed to get injections for the lungs before she was born – two weeks before, they gave the injections. Hannah was born, but she was born by emergency Caesarean which was planned – but obviously planned for a bit later because she wasn't due just then. So it ended up being an emergency and they took her straight to special baby care where she was vented for 51 days.

She came off the ventilator. She went to various different hospitals for long lines [a feeding tube system for babies who are unable to swallow] and... Mainly just long lines because her access was so poor. She kept knocking it off with her hands and then she had jaundice from three days old and they did the lights – the ultraviolet lights which they can do after 10 days – and they did it, but the yellowness never really went. We did think nothing of it. We thought it would go in time, which it didn't and she got more yellow and more yellow.

She was in special baby care for four months and then moved next door to the children's ward for another four months and then we went home, and then we came back about two weeks later because she'd become poorly and was transferred to Birmingham Children's Hospital where we'd actually asked to go there to meet a neurologist.

We went to Birmingham Children's Hospital to meet a neurologist who we'd met at a conference on myotonic dystrophy and we spoke to her about Hannah and she was interested in meeting her – so we had to go there and because she knew about the jaundice, she got a hepatologist involved as well and so we were then under the care of the Birmingham hepatologist and neurologist. The neurologist has now moved to Great Ormond Street, but she still sees Hannah – so it is that continuity of care thing, otherwise we'll just have to explain everything over and over again.

She's still under the hepatologist and the surgeons now at Birmingham Children's Hospital. Her liver disease seemed to get worse. Her bilirubin score kept getting higher and higher and it got to about a thousand repeats per... I don't know what the medical thing is for that – but she had the repeats of the bilirubin and they took us in to assess her for a liver transplant, but they waited until she was four because she was so poorly, they didn't want to... They just wanted to make sure that she would be healthy enough to have a transplant.

So they took us in to do the tests for it. She became quite poorly while she was there and in the end they said, 'No, you can't go home unless she gets a transplant'. She turned green towards the end, and we were put in touch with a hospice in Birmingham and we were referred to them. We were going to them on the 29th of November 2012, but two days previous, they rang. We stayed close by in a room in a guest house and they rang us and said, 'We've got a liver for Hannah', and she went down the next morning and got the liver and now she's seven, and she's like a different child.

Previously she never ate; she never talked; she never did a lot, really. She walked when she was three – so she's very delayed in everything. Her speech is not great; she does Makaton signing a lot and she's got an NG tube [a feeding tube through the nose] because where her liver and her spleen are so enlarged, they could never put a PEG [a feeding tube through the stomach] in or any of those sorts of things – so she's been NG-fed most of her life although now she's eating, but she can't manage fluids, so overnight (because it's overnight feeding) she has water and some feed to top up the amount of calories she gets and she's doing well. She goes to school; she goes to Brownies – she just started Brownies; and she's very demanding and very annoying (sometimes); bossy – yeah, she's doing well.

So, you talked a bit about when you were tested. You talked a bit about your sister. Can you say a bit about your other sister – Lorraine?

My sister Lorraine is the middle child and she's always been horrible. She was always a bit of a tearaway. She wasn't a good child, but being six years older than me, I didn't see a lot of her when I was a child. She'd go out playing with her friends, but she's nice now. She's better now. She's 42 and lives with her husband and her two sons and two dogs not too far away – about half an hour's drive. She has never really worked. She was not a good child when she was young and she left school early. She didn't do any exams at all because she hated school. She did some work in nursing homes which she actually really enjoyed, and she did some dental nursing after she left school, but she didn't stick at it for very long and now she goes around cleaning some houses, but she only does it one day a week as far as I'm aware. But she cleans her house all the time. She's always cleaning. She's alright. She doesn't really do much else, to be honest.

OK, and say a wee bit about your nephews – how old they are and if they're still at home.

Both my nephews still live at home. One will be 21 next month and he's hardly ever home, but we see him on birthdays and special occasions, really. He works close by. He used to work in London and had to commute quite often, but now he works closer to home.

My other nephew, he's 19 and he's... I'm not sure what he's doing. I thought he was becoming an electrician or plumber, but I've been told recently he's a mechanic

– so, I'm not too sure what he's doing, but he goes out a lot. He comes home; has dinner; and goes out again. I don't see them regularly because they live that bit further distance and I'm busy with family as well. We don't see each other that often. It's mostly occasions like birthdays and Christmas and that sort of thing.

Within your family set-up, is there one sister that seems more responsible than others? You know, if decisions have to be made in that kind of family way?

In terms of bossiness and responsibility in the family, I'd say that my sister, Lorraine, the middle sister, is more responsible for decision-making and getting things sorted out and planning things because she's quite bossy and she gets things done. She doesn't like to mess about. She's more like, 'Let's get it done now' rather than, 'Oh, I don't know...'

I'm not very decisive, and my sister is, like 'Oh, whatever'. So, Lorraine is the responsible one.

Could you describe yourself a little bit, just in terms of... You're at that age – you've had your family; how you feel in relation to where you were when you were like a teenager. Do you think you've changed since that time?

About myself, when I was younger – when I was sort of a teenage age, I got problems with my hands. I had really cold hands, and I had problems with grip, but we didn't think there was anything wrong, at that point. We just thought it was... Actually, my mum put it down to poor circulation, so we never thought nothing of it. We just got on, basically, how it went.

And then, as I got older, I developed more pains and that was put down to growing pains, but we never really thought all that much about it. I had very, very, bad period pains which was normal because my mum had very bad period pains, so we just thought it was a normal thing, but I used to actually vomit and be really, really ill. I'm not now, but previous to having children, I was very ill with periods but I was never really ill with anything else. I've had colds and things like that, but nothing out of the ordinary, and my milestones were all normal. I was not delayed in that aspect of things.

I struggled at school with some subjects, but not in a way that I needed extra help. I was quite good at reading and writing – just not very good with maths and languages. I did all my schooling, and then I went into work, and I worked with children for a while; I worked in a pub; and I worked at the hospital where I looked after adults, and then I went on long-term sick because I was getting so much pain in my legs, I could hardly move – so I went on long-term sick and then, in the end, I got medical retirement because of the pains and because my manager was not being very sympathetic to my needs and so I've been retired for seven years. I was actually pregnant when I retired.

Now I'm at home, but I don't do an awful lot now.

But you have the kids.

I have two children. I have Joseph who is 15, and he's doing his exams soon. He does fencing after school practically every day and he's doing quite well because he is nineteenth on the ranking for England. He enjoys that. It's quite tiring for us, though.

My daughter who is seven is at school – a special school, which she likes going to, and she seems to come home happy and tired, most of the time. She might have a sleep, but we have to wake her up before she goes to sleep otherwise she won't sleep at night-time, and she goes to Brownies. She just started Brownies once a week which, at the moment, she's not sure about whether she enjoys it or not, but she likes everything. She loves animals; she loves horses; she wants a horse.

How would you describe your health now – your health, not your family's?

My health is OK. I don't... I take tablets to keep myself awake. I take Modafinil, but that's the only tablets I take.

I've just been recently told I need a pacemaker which will probably need to be done sometime after May. Stephanie has already got a pacemaker. She's had it about a year.

But other than that, I have yearly ECGs, echoes, and I have to have yearly diabetic tests and thyroid, but other than that I am quite well. I have a wheelchair and I have a stick for when my legs are very painful because I suffer more pain in my legs and ankles than pain anywhere else, but I get very, very tired – even taking the tablets, I still feel very tired. I can get up after a full night's sleep, and want to go to sleep again because I'm so tired – so, generally, I do.

The tablets are alright if I'm busy. If I'm doing something, if I'm going out shopping or going out or meeting somebody or whatever, then the tablets help, but if I'm not doing a lot, they don't help. I just sort of sit down and have a cup of tea at home and I might fall asleep.

If you could change one thing – if there was a magic wand or pill that will take one of the major symptoms away for you, what would you change?

I'd rather there not be any myotonic dystrophy – but if I have to change one of the things, I think it would have to be the sleepiness because that really affects me quite badly and it affects my life and it affects my husband's life because then he has to drive a lot of the time. I do drive, but I get tired very quickly so, quite often, he does a lot of the driving – so, if I could be less tired, less sleepy, and probably get better quality sleep, yeah, I think that would be the one thing I would change.

When you think of these boxes that you get sometimes in forms about what age you are, how would you describe yourself – as young, middle-aged, older? Clearly you're not retired.

I'd tick myself as retired – on the boxes when they say, 'What is your profession?', I tick 'retired'. But if they did an age thing, I would probably put... I don't know if I'd put middle-aged because I don't feel young-young, but I don't know, really, to be honest – because normally they put ages so I normally tick the age box.

When you and your sisters were growing up, did you wear each other's clothes? Have you got any stories about falling out? There's quite an age difference between you.

I never borrowed any of my sisters' clothes, and they never borrowed any of mine but my two sisters – they used to swap clothes all the time, and they've got the same size feet as well, where I've got bigger feet now. My sister has borrowed my boots before, but she had to put socks in the end so that they weren't so big, but we never really shared things.

I used to get toys that were my sisters' and clothes that were my sisters' when they'd grown out of them, and I was still growing up – but other than that we never shared clothes.

When you got your diagnosis, did it explain anything for you? How did it make you feel when you found out about myotonic dystrophy?

Yes, when I was diagnosed, and after I'd spoken to the genetic counsellor and she'd explained a lot of things, a lot of things that had happened previously, we'd look back and think, 'Oh, that's probably connected to the myotonic dystrophy', and then speaking to the genetic counsellor again, or the neurologist, we'd say things like, 'Oh, this happened so many years ago. Do you think that's connected?' and they would say, 'It quite possibly is connected', but they can never say for sure whether it would be because you never really know – so, a lot of pain I was getting and the periods, they said that was probably connected. And the sleepiness – I was quite a tired child as well.

And then with my dad. My dad never showed any symptoms, but after I was diagnosed and he was tested, we found out that there are so many different things that are associated with myotonic dystrophy – we found like the cataracts with my dad and diabetes and that's how we thought my nan had it.

My sister had her thyroid removed, but we didn't know – that was before we knew about myotonic dystrophy, so we found out that was involved and we found out that the gallbladder was involved as well because she's had that removed as well. Anaesthetics were... We didn't know about anaesthetics – so, in the end the anaesthetics that were given, we didn't know at the time, could affect you...

Just go back to your sisters again. When I was growing up, we were always told, my sister and I, that we are very alike and I wonder if that's something that ever happened in your life?

I never really thought I looked like my sisters at all. My older sister, Stephanie – we have been told that we look like twins and been mistaken for twins before but I'm not sure... but that was, we were told we're alike.

I don't think I look like Lorraine at all although, when you look at pictures, you can see similarities – but she's always been a bit different, whereas me and my sister, we've got the same colour hair, same facial features and we both wear glasses – so we look a lot alike, but other than that, nothing else.

How about fashion? Have you had to change the type of shoe that you wear? Do you like shoes? Do you have a big collection of shoes?

I'm not a fashionable person. I don't go with the fashions of the day. Lorraine does; Stephanie tries to – but I never was into heels all that much, but I did have some heels and I've had to stop wearing heels because they hurt my feet too much and my ankles swell a lot of the time. I tend to wear flat shoes or trainers a lot of the time. I've got boots which I've had to buy that have elastic in them because my calves are quite big due to the muscles in them. Some people say I have footballer's legs, but quite often I tend to just wear trainers a lot of the time.

Do you wear splints at all?

No, I don't have splints. My sister, Stephanie, has splints – but they're more like just foot ones, but she doesn't have the swelling that I have. I don't know why my ankles swell, and they're not sure why they swell either.

How do you feel about the future for you and your family? Is it something that you're optimistic about, or do you have any worries about that?

I do worry about the future – what it holds for the whole of our family. My son – he's fine. He's not been tested, so we don't know whether he's got any type of problems, but he doesn't seem to...

I worry that I'll die earlier than, say, any normal person, and my daughter as well. We've been told that she will die earlier than normal but, obviously, that might change – but other than that, I'm hopeful that the pain won't get any worse and I'm hopeful that my daughter will progress and maybe...

Actually, she was on a drugs trial, but she was taken off because of tumours in children under 12, but she's doing that well, I'd like her to be alive at a time where there might be a cure, or there might be something that can help stop it progressing anymore – but I'm not sure that will be available for me because I might be too old by then.

But I'm hopeful that we'll have a good life – but, no, nothing not to look forward to, but I've got nothing planned, in a way. It's just sort of get on with day-to-day life and do whatever.

Do you ever think about ageing treatments? Is that something you discuss with your sisters?

I don't really discuss any regimes or beauty treatments or any sort of ageing things with my sisters. I do joke that they're older than me. Lorraine is probably more vain that Stephanie and she pulls out any grey hairs you might see and she said that she found one in my sister's hair the other day.

I used to dye my hair a lot when I was younger because that was a school in-fashion thing, but I haven't dyed my hair for a long time. But I did say, or I have said to, probably, more my husband, that I'd dye my hair if I go grey.

But I don't follow any beauty regimes or anything. I don't even use moisturiser which I know, I probably should, all the time – but I only use it occasionally if I feel like it but, other than that, I don't go in for the beauty things.

I did joke that I wanted Botox here [*indicating frown lines*] because I said that was a bit squidgy and I said I'd want Botox here, but I probably wouldn't really go for it. And I wouldn't go for plastic surgery or anything like that – but other than that, I'm not really a fashion person. I never talk about fashion or beauty or ageing.

I don't feel that old at the moment. I don't feel too old. I don't feel young because I've got a 15-year-old son – that makes me feel a bit older, but I don't feel *old*.

Is there anything you would like to say? Is there anything you feel scientists or doctors don't recognize about what it's like to have myotonic dystrophy?

I think there are some doctors that don't listen – especially when you've got a child with the condition. A lot of doctors... There are some good doctors out there that listen and take into account what you've said, but there are some doctors that don't listen, and ignore everything you've told them and they get everything wrong and I think listening to the patients and the parents is a good thing because they know more than the doctors normally know. It's not that common a condition that they know about that much – but, yes, so probably just listening and accepting that we know more than they do.

That's all, really. Everyone else is generally OK. We have problems with feeding and swallowing things because they... It's like they've got a rule book that they have to follow, and Hannah doesn't follow that. We fight with them. Other than that, they're generally ... Most people do listen to you, and we've been quite lucky with the doctors we've had. That's it, really.

Stephanie, 44

Age at diagnosis: 33
Essex, 15 May 2015

Jackie: *If you can tell me your name, and how old you are.*

Stephanie: My name is Stephanie and I'm 44 years old.

And tell me where, and when, you were born, and where you live now.

I was born 1970 in Essex and I live in Basildon.

If you could tell me if you have any siblings, and what their names are.

I've got two sisters called Lorraine and Michelle.

And then, if you can just say a sentence of how you sit within your family – the position within your family.

I've got [two] sisters. I'm the oldest. Lorraine is my middle sister – well she's the middle one, and Michelle is the youngest.

Can you tell me a wee bit about your childhood – particularly when you went to school? If you liked school, if you didn't, whether you stayed on.

I went to Northlands Primary School where both my sisters went as well, then we went to a comprehensive school. I quite liked school, but I didn't stay on. I went to college and redid some exams I didn't do very well in, and then I went to work.

Could you just talk about what you did? You then went back to that college – what you studied when you went to college, and what you went to work as.

I went to college and studied mainly science subjects and then I went on and got a job in a chemist which wasn't very good, so I left there and I got a job at Yardley's Cosmetics and I worked in the biology department in there, and then went on to research and development of new cosmetics and stuff.

Can you say a bit more about that? What kind of research? What were you looking at?

When I worked at Yardley's, originally I was in microbiology and we had to test the surfaces to make sure they were going to keep any bugs at bay and then I went into Research and Development and it was sorting out new ingredients for the make-up to make it better and last longer.

Tell me what you do now.

I don't do nothing else. I stay at home.

Maybe talk about that period when you stopped working – is there a reason behind it?

Well, no, it's just – well, really, I got signed off because I couldn't do my job because I lost my job. I was doing a driving job before. I lost the job because I had sleep apnoea and driving – you can't do that. You can fall asleep at the wheel – so I left that, and then I couldn't really get a job and then I had to go onto a disability living allowance and now I have unemployment support allowance as well which I got through my consultant at London because they wouldn't give it to me to start with because they said I was fit to work.

So you had to go through all the assessments for that?

Yeah. Well, I had to go for a medical which I failed, but then I re-applied and they said I had to go for a medical, but then they wrote to me again and said that they had the information they needed and that I didn't have to go for a medical.

I'm going to ask you a bit about your home life here, whether you're married, and if you have any children or pets or anything.

I'm married to Ian. We've been married for four years, but we've been together 13 years. I've got a stepdaughter who is 26, and a stepson who is 22 and I've got one dog and one cat.

Next, if we could talk a bit about myotonic dystrophy – so, what you know about it and how you found out that you have it – the process that you went through to be diagnosed.

Myotonic dystrophy – I know a fair bit about it. I know it affects your muscles and your heart – well, your heart is a muscle. Well, that's all, really. I know there are all different things that can happen.

I've got it on me. I've got a card – an alert card and it shows you on there what can happen and that.

Maybe we'll start the other way round – if you don't mind talking about how you came to be diagnosed with myotonic dystrophy and whether that diagnosis explained anything. You know, some people – my sister was always really tired and then, when she got diagnosed it kind of joined up. You know, it made sense, in a way.

I was diagnosed when my sister had a baby which, bless it, was floppy baby syndrome. They actually tested her and found out it was myotonic dystrophy that she had and so my sister was tested, and I was tested, which were both positive.

How did that feel at the time?

Well, I just didn't want to believe it, really. Sorry – I didn't want to believe that I actually had myotonic dystrophy.

Do you feel it did explain anything about your previous health history when you got the diagnosis?

Not really. The diagnosis, I think, it didn't really affect me at all until then, and then I did notice things, but I never noticed any other things that happened before, like I had spasms in my hands for quite a long time – for years – which was probably one of the symptoms.

My legs were OK, but now they get tired a lot as well, and pains, and falling apart – and, also, I've got a pacemaker fitted due to the myotonic dystrophy.

OK, so have you noticed any deterioration in your symptoms more recently? Have you noticed anything that has changed?

Only the deterioration, really, in my legs with the aches and pains and I think it's taking longer to heal when I tear a ligament or walking takes longer to recover from after I've been out. I don't walk a lot now, anyway.

You talked a bit about some injuries that you've had. Can you talk more about them? Have you fallen recently – things like that?

I haven't fallen recently – not for quite a while, but that's probably because I use a stick most of the time or go in my wheelchair. The most serious injury was when I was on holiday and I fell down a step and broke three bones in one foot and tore the ligaments in the other one, which was quite painful.

Thinking about your two sisters. Your younger sister was diagnosed with it at the same time as you. Her daughter has quite marked symptoms, but have you noticed any of the symptoms becoming clear in your sister as well?

No, I can't see any symptoms that are clear in my sister. I think it's not as bad as what I've got. Even though they say that hers is onset earlier than mine – but that might be just because I was diagnosed later – at a later age, like, because I'm eight years older than Michelle.

Could you describe your sister, Michelle, to me as though I haven't yet met her? Just what she looks like, and what her personality is like?

My sister, Michelle, she's a fair bit taller than me. She's about 5 foot 6, and she's got long, thick, dark brown hair and brown eyes, she's quite big-built and she's got bigger feet than me as well.

She is quite quiet, but I don't know whether that's just with me, or whether she's just a quiet person, but she is of medium intelligence, and she's quite caring with everybody – especially with me.

If you see her, does she help out? Does she sort of look after you, in a way?

She sends me godmother cards from my niece and nephew because I've got no children.

So, going on to your sister, Lorraine, could you describe her for me?

My sister, Lorraine, she's taller than me as well, and she's very slim, and she's got long, quite thick, dark hair, brown eyes (we are all alike like that), and she's quite an outgoing person. She's quite caring as well. Yeah, she's very family-orientated with her two boys and all that.

I don't see her a lot, really – but we are quite close when we get together – but me and my sister, Lorraine, we sort of gang up on Michelle because we grew up more together than my sister. We both moved out when she was still growing up.

So you were closer to Lorraine?

Yeah.

So your sister's got two boys that she seems to run after a lot. Do you want to say a bit about her family life and how she runs her house?

My sister, Lorraine, is very house-proud and I think she's got OCD as well. She's very... stressed. She gets very stressed, trying to run the house, keep it tidy, because Richard, her husband, is a plasterer and he's always decorating and she's got dust everywhere and her OCD makes it worse and then she's got the two boys who run her ragged. She has to run them about – well, not so much now because they've both got cars, but she's usually picking them up or taking their girlfriends home and things like that and she also tidies their rooms because they're boys and they don't do that and it just gets her goat.

Do you think that you look alike? Are there any family resemblances between the three of you?

Family resemblances... I think me and my sisters take after my dad. We've all got dark hair, and we've all got brown eyes, whereas my mum's got white hair with blue eyes – so she's quite fair, but we all take after my dad, but I think all my sisters look quite alike, I think.

We have talked about your symptoms. Is there anything you want to talk about or anything that particularly bothers you?

My symptoms – I think the worst thing is my legs because I can't do things I used to do, like walk round shops and things like that and I feel like it's a burden on my husband. He's got to do more things for me that I could before, like opening jars and things like that because I can't do things because he says to me, 'Oh, yeah, I should get a carer's allowance if I'm doing this for you', and things like that. But mainly it's my legs.

If you could say a bit about your appearance. How, if at all, has myotonic dystrophy affected your appearance?

The only thing that's affected my appearance is my hair because I've got not much hair which they said is probably one of the symptoms as well – that's why I wear a hat or a bandana or whatever just to cover it because people stare. I've got wigs I wear sometimes as well because it looks nicer when you're going out to wear a wig – but apart from that, I feel happy because I've lost weight as well which made me look better and I think that helps a bit with the weight on my legs where I've lost a bit of weight.

So it helped your mobility? You feel more able?

No, just the same, but... I don't really know.

Thinking about your appearance now, compared to 10 years ago, do you think you look any different to how you did 10 years ago?

I don't think I look a lot different to then, but I've got a few more wrinkles. I had more hair then, which was better, and I was a bit fatter then – but apart from that, I don't think I've changed a lot.

Do you see any differences in your sisters? Do you see them ageing at all, now that you're in your forties?

I don't see Michelle ageing, but I always think of her as being much younger than what she is. Lorraine looks like she's aged. She's got crow's feet and she's always complaining about all the wrinkles around her neck but, apart from that, I don't think they've really changed. They're just the same.

What about your parents? Have you noticed them ageing?

My mum has aged. Her skin looks a bit older and she's got white hair – but she's had white hair for quite a long time but she's quite able, whereas my dad, he's got myotonic dystrophy anyway, and he's going to be 70 this year and he looks like he's getting old now. I don't think he can walk as well as what he did. He gets tired a bit quicker, and he's not as agile – but apart from that, they're alright.

If you wouldn't mind talking about what it was like growing up with your sisters. I know you're the eldest, but did you ever borrow each other's clothes, or go out together? I know there's a bit of an age gap with your younger sister – but if you can talk about those relationships before you all left home.

Me and Lorraine, we used to fight a lot – badly, and break things and stuff. She'd break my dolls and throw my make-up across the room even though we were quite good friends – still. We didn't like sharing a room and she used to nick my clothes and I used to really hate that because it used to be my new clothes, she'd go and put them on before I'd even worn them. And she left home when she was 16 – so we got on better once she'd left home.

My other sister, Michelle... Well, we used to just gang up on her – me and Lorraine – on Michelle and pick on her. I think she was a bit lonely because we didn't really have much to do with her because we used to be out with our friends. I used to take her out, like take her swimming. I taught her to swim. I took her swimming and stuff and... I got paid some money – pocket money and that because she was that much younger. That's all, really.

Michelle has been through quite a journey with her daughter. Can you talk about your relationship to Michelle's children, as their godmother – how you see her family, because she's younger than you?

I'm quite close to them, I think, because I go there a lot and they got used to seeing me. Joseph is at that age now – he's 16, so he's in his room with his Xbox and he's studying for his exams now, but he's very nice, very intelligent, and he does fencing which he enjoys – so, I didn't really do much with him. I used to play with him when he was really little, and read stories and that.

But Hannah... Everyone says she looks like me! She's been through so much in her seven years. [*Upset*]

I went to see her. She was in Birmingham when she had her transplant. I went up to see her. I didn't know if she was going to survive or anything like that. She loves cuddles.

So you're close to her?

Hmmm.

So say a bit about some of the things she is able to do now, since she's had her transplant and her health has improved.

Hannah had a transplant two and a half years ago, and since she's had that, her health has increased dramatically. She's grown taller, her hair has grown and it's nice and quite thick. She can do more running about and playing and that, and she is talking – not a lot, but more than she was doing. She wasn't doing any, and she can count and she can do a little bit of writing – not a lot of writing, but she can do some writing, and she's eating lots of food whereas she was tube-fed and she is off food through her tube now and just liquids which is really good. She loves her food – especially spicy food and I think she enjoys going to school and the Brownies. I think she likes the Brownies. She can mix with more children. Yeah, she's just a very happy, healthy-looking, child now.

Do you want to talk at all about when you had your child?

I had my daughter 11 weeks premature. I had polyhydramnios, which is the excess fluids which build up, which meant I went into premature labour. We didn't know there was anything actually wrong with my daughter. They just said it was oxygen because it was so quick. She was born at home – so, that wasn't the best because she wasn't breathing or nothing, and they took her in and she was in intensive care just for the day because she wasn't responding and they decided that it was best if she had the machine turned off – but looking back at it now, she was a lot like my sister's daughter, Holly, that died, and we really think that she had it as well, although we didn't know at the time because we didn't know it was in the family when I had my daughter.

And do you feel that joined up with your sister losing her baby as well, when you did find out – did it give you any sense of understanding of what happened?

Yes. It didn't make me feel any better, but now I know it wasn't anything I did or anything like that.

Now that you have all found out about myotonic dystrophy, what have been the effects on your whole family – even the people who are not affected personally? Maybe also think about your mum who doesn't have it, but is clearly very affected by it because so many people in her family do have it.

I don't think it affects them very much. I think my mum worries about us and whether we're doing too much. If I go out with my mum she always says, 'Hold on', just in case I fall over, but I think she's quite easy-going with it, and my dad's not badly affected – so that's alright. Ian is alright.

My sister is the one that's not been tested. She's fine. She just says she'll take me out in my wheelchair. Because my mum goes, 'Do you want to do that?' to start with, because she thought it's just making us worse which is, it's not making us worse, but I think my sister tends to use her wheelchair more than me because I think, if you don't use it, you lose it – basically – but I try to keep walking as much as I can just to keep my muscles going a bit – but that's going off the track a bit really, isn't it? But Ian says, 'Oh, you're in pain today, let's take your wheelchair', because we went down to town and I was really slow so we went down Lakeside. (You know Lakeside? Lakeside is a big shopping centre like Bluewater.) He took the wheelchair because he said I'm too slow. But he helps a lot. He does a lot in the house as well.

And does your sister's husband help her also?

What, Michelle's husband? Not really. He does a lot of running around rather than Michelle, [or he's got a lot of daddy skills] so he does a lot of the cuddles and that more than Michelle but, yes, he goes out with them and his son – he takes him fencing and stuff like that. Other than that, he doesn't really do much in the house. It's not very tidy.

But my sister, she does everything, so she waits on her husband. He is out at work quite a lot doing his plastering and that. Yeah, I think my husband is the best – but I'm supposed to think that!

Do you ever wonder what your sister – particularly Michelle – would be like if she didn't have myotonic dystrophy? Is that something you ever thought about?

I don't think Michelle would be a lot different because she's always been a bit strange. I don't know because she don't do much indoors. She sort of sits on the computer most of the time – but I don't know whether... Maybe she would be working if she was... And if she didn't have Hannah. Because if she didn't have it, and she had Hannah, Hannah wouldn't have had myotonic – would she? I think she might have had a job, maybe, and Simon would work as well because neither of them work. They stay at home because Simon is the carer, but I don't think she'd be a lot different to what she is now. She might be a bit tidier, but not a lot, no. I mean, I'm not the most tidiest of persons. I'm not as tidy as Lorraine, but I'm tidier than Michelle.

Thinking less about the last 10 years, and more about your future, what are your thoughts on ageing? Are you looking forward to being more mature; is there anything you worry about...? Your sister has quite clear views on her thoughts on ageing. She wasn't happy about it at all.

I'm not looking forward to ageing because I know this is progressive and I know it's going to get worse and at one stage I might not be able to walk at all which worries me, and I might have, like, bladder problems and things like that because that's not the best at the moment anyway. I'm on tablets and that.

I worry about my heart. I don't know whether the pacemaker is going to help it stay as good as what it is now – so I worry about that as well. And my eyes because I've got cataracts – very early-stage cataracts, but they can do something about that. So, yes, I'm not really looking forward to it progressing... But it depends... It might be slow progressive, but I don't know.

What do you hope for in terms of research? If you had a magic wand, what would you hope that they could address with treatments?

Treatments such as stem cell research. Maybe something they can inject into your muscles to help them. I don't know what it is. It's deterioration of the muscle, isn't it? Maybe something to do with that and if they can look at something that will stop it progressing any more than it already is and maybe make it less severe than it is when you first get it – so that's really all.

Yes, there are some interim treatments, like you had the breathing machine and there's physical things you can get done with your eyes – some medication you can take and some symptomatic treatments available. Is that something that you would consider? Does your consultant keep you up to date with the treatments that are available?

Well, no, he doesn't give me any treatments, but I think that's because a lot of the stuff wasn't suitable with the heart problem, but since I've had my pacemaker, I probably could have something like Minoxidil – that's for my hair. There's a couple of things my sister's had but if I've got a pain, I just take some painkillers, really, at the moment. If I don't need them, it's better not to have them right now – more when I do really need them, I think.

Thinking more about ageing, would you consider any sort of physical treatments like Botox or plastic surgery or anything like that?

I wouldn't mind a boob job... A lift! And I'd like a tummy tuck as well. Apart from that, I mean, I'm alright as I am, I think. I don't think I want to do that – I mean, surgery is not advisable when you've got myotonic dystrophy anyway because of the anaesthetic and that – so, I'll just stay with what I've got, I think.

You said you're going out tonight. What's your favourite look for going out?

Usually smart trousers and sometimes a dress, occasionally. I might wear jeans tonight, though. I've got a new pair of jeans. We're only going like to a pasta place for dinner. I mean, I don't like pasta, but they do pizzas, but I don't eat pizzas. It's so fattening, but it's only once – so...

I asked everyone about shoes – whether you like them, what kind you like to wear, and if your fashion sense has changed.

I like shoes, but I'm very restricted now what I can wear. I can't wear heels. I can walk on a wedge, but I can't wear heels, I just fall over. I just can't balance. The balance is the worst, so it is flattish shoes, boots, and trainers, really. I've got a few pairs of sandals now which are more feminine because a lot of the shoes you buy aren't very feminine and they're not like what you can wear out – but I've got a few pairs now. Not like Lorraine. Lorraine's got a pair for every outfit and she actually had a room which was her shoe room – tons of shoes, tons of handbags, and loads of boots. She's given me a couple of pairs of shoes as well which she bought because she liked them, but she's not the same size as me and they didn't fit her, and fitted me – so I got her shoes!

But, yeah – I wish I could wear heels... I've still got some heels, hoping I can wear them, but I just look at them and don't wear them because I'm just going to not walk very well – worse than normal.

Extracts of this film will be shown at conferences with scientists. Is there anything you feel you'd like to say to the world of science and medicine? When you go to see a consultant, you have such a short visit – so, is there anything you feel you want them to consider?

I think I'm quite happy because I actually go as an inpatient now as well.

I go up to the Neurology and Neurosurgery Hospital in London at Queen Square to see my consultant and they've got an inpatient area where I stayed in for four days last year and they want me to go every year as an assessment to re-assess everything each year like an MOT [motor vehicle annual assessment] which I think is good – but that's all I really want to say. I mean, I'm happy with my consultant – well, all of them, because I've got one for each thing – one for my breathing which, he is good because I've stopped that now. I just have to have a study once a year. I don't see my heart consultant anymore but I think I'm due to see him because I haven't seen him since I've had a pacemaker. But I have checkups every year at Basildon Hospital to check it's all working, and my main consultant I see once a year which is just a brief update on the past year, really. And I've got a neurology doctor now as well which, I've only seen him once, but I think I'm due to see him sometime this year as well. That's all, really.

I ask everyone to talk about their fashion sense when they were growing up. You talked about shoes, but maybe if you can recall your best and worst fashion moments as you've matured?

I'm trying to think of something that was really bad... We had these horrible long dresses – it was purple, with all these big checks, and it was long. We wore it for these photos because we really liked it – me and my sister, and we used to wear those Jesus sandals, and we used to love them – and it was horrible, really. I suppose, it was in the seventies, and we liked that sort of thing but it was horrible when you look at the photos now.

Yes, so the last bit is to ask if you've ever been to any support group meetings, or family group meetings, or whether you've ever met anyone who has myotonic dystrophy who is not in your direct family?

I've never met anyone else out of my family with myotonic dystrophy. I've never been to a support group, but I've been to a seminar which was a lot of all these consultants and specialists for a whole day – a hotel for the day, with my sister, Michelle, which was quite interesting – some of it.

So you've never met anyone else? You've never had any contact with other families with it?

No.

Is it deliberate, or is it just something that hasn't happened?

No, it just hasn't happened. I haven't met anyone and I haven't been to support groups. I think there is one in Essex, but I don't know where it is, and I think it's quite far away and I'm happy as I am... But I don't know.

There's a forum on Facebook which I go on there, but I only read it; I don't post anything myself. I just read, because I'm interested in what's happening to them people – a lot of them in America as well.

My sister finds the Internet quite daunting. Do you?

Some of the stuff the Americans say about myotonic dystrophy is quite strange, but I feel sorry for some of them because they have to pay and sometimes they won't be funded and things like that, and it must be really difficult over there to get the diagnosis. I mean, people moan about the NHS [National Health Service], but I mean, we're quite lucky in that respect.

Is *there anything else you want to say? Your thoughts on how your family has changed since you found out about myotonic dystrophy and whether you think the discovery has changed any of you?*

Not really. I don't think it's changed a lot. I don't think a lot of my family life has changed. It's not any worse, but it's not any better, really. It's just a little bit harder with the disabilities and stuff.

I'm not very good at talking. I don't talk like my sisters. I'm more quiet than them, I think.

Michelle talks a lot.

I think Lorraine talks more.

Lorraine said the least.

It's probably because she doesn't know so much about the myotonic.

Michelle spoke a lot about Hannah.

She knows loads as well.

Like the history of it...

I think her husband Simon knows more than some of the doctors with the stuff he knows.

Some people do that. Some people want to know everything while others don't.

When we went to the seminar, my husband came with us. He didn't really want to go, but I wanted him to go so that he understands it a bit more than what he already did because I don't think he really understood much of it.

Does he relate what you've got to what Hannah has?

A little bit – not a lot, because she's much... Well, she was much worse. She's a bit better now – but he says I'm worse than Michelle. He says I'm much worse than Michelle.

But that is because you're a bit older. When she's your age...

She might be worse than me.

Or just the same.

I think they don't affect boys as bad as it does girls, apparently.

I think when a man passes it on, it doesn't multiply as severely as when a mother passes it on.

My dad passed it on to us.

It's really common. The way your family has found out – the father passing it on to the daughters or sons, but they don't know, and then the daughters have very sick babies – congenital – and then that is how they trace it back. That's the most common way that people find out.

Yes, it came from my nan to my dad. We only realized it probably was my nan because she had all these diabetes and all that so they think she passed it to my dad and she passed it to my aunt, but my aunt probably hasn't got any affected children. She's got three children. One might have it – well, she looks like she might.

And do they have children?

The two that most definitely haven't got it, have got children, but the middle one – she hasn't got children, and she doesn't want children, but her husband's got two children. But I would have liked one.

Do you miss not having kids?

I think so. I've got Ian's kids – but it's not the same. When I got together with Ian, John was 10 and Sarah was 13 – so they weren't babies. That's probably why I see Hannah and that quite a lot.

Lorraine, 41

Essex, 17 March 2015

Jackie: *Let's start with your name and how old you are.*

Lorraine: My name is Lorraine and I am 41.

Can you tell me about where you grew up – where you went to school, a bit about your childhood?

I grew up in Basildon, in Essex. I went to comprehensive school [a state secondary school]. I left school when I could because I never enjoyed it. I enjoyed going out with friends – a bit of a free spirit. I met my husband when I was 17 and moved out of my house. I've been married for 16 years. We've been together for 25 years now.

Tell me a bit about your own family life.

I became a mum at 21 and again at 22 – I got two boys. They're grown-up. My eldest son Edward is a very laid-back boy; very popular. He's trained to be a quantity surveyor.

My second son is a troublesome son. He's a trainee mechanic; loves cars. I haven't got a clue about cars, so I try and be interested in what he does – about both boys.

Tell me a bit about your two sisters – your position in the family and what it was like growing up with two sisters.

I'm a sister of three. I'm the middle child. I have an older sister which, I feel, I've probably got more in common with – most probably because we're closer in age and we grew up more together, like playing out and everything. We do have a laugh together.

I have a younger sister where there's a bigger age gap. She probably feels like an only child. We do get on quite well, I think, as we got older, but I am the odd sister out. I'm quite outspoken. They probably will say that as well.

How do you see your sisters? If they see you as outspoken? How would you describe them?

My older sister – she's quiet. Quiet and funny. She doesn't get a joke and then she does after 10 minutes and laughs. She's quite a nice and sensitive person, I think. She doesn't really say anything. She doesn't smile much, but I think, looking back now, I think because of the muscular dystrophy, it has affected her from smiling because we're always saying, 'Smile!'

My youngest sister – she talks forever. She's always got something to say. She loves her kids – dotes on them. A lot of her time is taken up with her youngest one because of all her illnesses. It's horrible to say that I've not really got on as well with my younger sister as with my older sister, but as I'm getting older, I feel that I do a bit more.

How would you describe your older sister to me?

My poor older sister has lost a lot of her hair which is quite sad, really. She's lost quite a bit of weight. She's a little bit shorter than what I am and rounder in the body. She looks like a serious-faced person all of the time because she doesn't... I think she hasn't got the muscles in her face probably to smile as much.

Do you see a family resemblance between the two of you?

I think if you look at my older sister, and her personality, she's probably more like my dad. They don't really show you their feelings.

I'm probably an emotional, outspoken person. Michelle – not so much. I think, probably because I don't spend so much time with Michelle... I don't spend much time with either sister. I think they tend to get on better; always have done. I think, probably because, now, because of the muscular [sic] dystrophy, they've both lost children so, I think, they're a lot closer together. They've got more in common in what they do and enjoy, where I'm... I won't say 'black sheep'; I'm just sort of the middle child and just get on with life and just do my own thing.

I'm very close to both my parents – now. I never did with my dad as a child. That's probably because we both have to have the last word, but I think I'm probably more like my mum in being sensitive. I think, probably, Michelle and Stephanie are more like my dad and he's the one who has the muscular dystrophy gene.

Do you feel you look like your sisters?

Because there's us three sisters, I am the middle child and very different, I do think we do look quite similar. I think my older and younger sister look maybe more alike, but I think we do have certain features that everybody can still see that we are sisters. I'm the more beautiful one! [*Laughing*] Don't put that in!

In terms of your parents, have you been told, 'You're so like your dad; you're so like your mum'? Is there a family resemblance?

I think us three sisters look more like my dad. I think it's probably because we're all dark. We've got the brown eyes, dark features, and probably short legs – and that's my dad's side of the family.

Can you say a bit of what you know, if anything, about myotonic dystrophy and how you found out – how it came into your life?

I don't know... well – hardly anything – about myotonic dystrophy, only that my sisters have got it.

We found out through my younger sister that had a baby and there was something wrong and that's how they found out it's dystrophy.

Previously to that, my elder sister lost a baby, but we thought it was through lack of oxygen. Now we know that it was probably through this myotonic dystrophy.

The only thing I really know about myotonic dystrophy is it affects all your muscles. I noticed that my sisters' hands – they can't do specific things. It is, I suppose, a muscle-wasting disorder and they're both very similar. I've noticed that they're both sort of sleepy people where I never sleep – but I don't really know too much, probably being ignorant of it because, I don't know, maybe because I haven't been tested for it and don't want to. I'm just sort of ignorant because I don't look into it in case, if I had a twinge or something, it could be, 'Have I got it?' and then it preys on you so I just play ignorant, I think, towards trying to know anything about it. I just know that it's to do with your muscles.

So it's a protection thing.

I think so.

It's often stated as a family illness. It affects lots of people in the family – even people who don't have it – a bit like your mum. She is somebody who doesn't have it, but two of her daughters do. So, maybe if you talk a bit about how myotonic dystrophy affects your wider family – if it's changed your family dynamic

Since we found out about myotonic dystrophy, I think my dad feels guilty because he found out that he's the carrier for it, but it's no one's fault because it's just one of those things. There's a name to it now. At least we know what it is. Since finding out that we've got it, we've realized that my dad's sister has got it worse than my dad because I don't think you'd realize that my dad's got it.

Now that the years have gone on, we can see a deterioration in both my sisters that has progressed a little bit more. I don't know if that's through them knowing they've got it. We're all supportive of them if they need us or have to go to the hospital. We're just there. I mean, everyone's got to get on with life.

You have decided not to be tested – is that because you feel there is no way you could have it, or because you feel it's just something you don't want to go near?

I've not been tested for it because I don't want to know if I have it. I feel that I don't have it because I don't show symptoms like my sisters have. Both my sisters will probably say that I haven't got it.

I don't really want to be tested because, if I do have it very mild, it ruins all your life insurance policies, I've been told through a relative that has been tested for it, and she has got it very mild, and it's ruined her sort of insurance, like life insurance policies and everything. I think if I did start – I don't know – deteriorating; if I felt that I had some kind of symptoms like my sisters have got, then I may get tested for it, but in the meantime I wouldn't get tested unless I showed signs of any symptoms like my sisters have got of weakness, sleepiness ... I don't have any of that.

How do you describe yourself?

I'm a family person. I speak to my mum twice a day, every day. I do everything for my boys which probably hasn't helped them because I run around after them even now that they're young men and should be doing stuff for themselves.

I enjoy going on holidays with my husband. We enjoy the kids being older now. I like going out shopping. I love to shop. I have two dogs. I go out walking with them. I enjoy moving... Well, I don't enjoy moving house, but I do enjoy having nice things, like most people do. I've probably just got a very nice, comfortable life and I'm very lucky.

You talked about how you feel lucky. Do you feel a sense of loss that your sisters and their children aren't able to do the things that you've been able to do?

I do feel lucky because I've got two healthy boys. I have an older sister that I feel quite sad for because she's not got any children and she would have been such a good mum. That's very sad. I do feel guilty that she's not been able to have children.

My younger sister – she's lucky and unlucky, really. She's got her oldest son that's well and then she's got a daughter that needs 24/7 care but she wouldn't be without her so, in a way, she's lucky that she's got two children where my older sister hasn't – so that's sad.

I suppose, as a child, I thought I was always [hard] done by because I'm the middle child, as they say, but really I'm the lucky child, really, because I've not got their illness. I can do anything and go anywhere where they can't do what I can do – just go up to London on a train. They have to organize what they do – so I am lucky, and I do feel guilty in a way – really.

You have a role within your family, though, because you are not affected. You didn't plan it. So, now, do you feel a sense that you are able to do things... Well, you do clearly feel that you are able to do things that they are not able to do. So, do you feel that that has changed your position in the family – you are now the steady one? Maybe an odd transition from having been the troublesome one that left home young?

Maybe you could reflect on that?

Because I live a little distance away from them, I'm not in their life as much. Really, we only sort of see each other if we're round our parents' house or birthdays, Christmas, because, I suppose, everybody has got their own busy lives going on. I'm always dashing here, there, and everywhere, like everybody else.

Their husbands are their supporters, and my parents are both always there running them round or helping out with my younger sister's children if one child needs picking up because they have a hospital appointment up in London.

They always know that I'm there if they need me, but at the moment I'm not needed, I don't think – but they do know I'm there for them.

Just talking about yourself now, your kids are quite grown-up now. Can you reflect on how you feel about ageing?

Getting older is not very pleasant. Getting older hasn't got any benefits as you start getting grey, getting wrinkles, but sometimes getting older does have benefits because your children are getting older, so you tend to get your life back to not having to run around after them, so you sort of tend to be able to go off and do your own thing, but I don't feel any older. My brain is not any older. My body is getting older, but my brain is not.

So thinking about fashion – you're the middle sister, so did you get passed-on clothes, did you borrow clothes, did you have an entirely different taste from your sisters?

As a child growing up, I did get hand-me-down clothes from my older sister, but because I was a different child, I didn't like anything that my older sister had, so I used get a right hump and throw a strop. But, yeah – I do like fashion, having nice things.

I'm lucky because I can afford to go out and buy nice stuff. My older sister gets my cast-offs now which she quite likes because she's always in my wardrobe having a look. But my younger sister, she's a bit taller and I don't have anything to fit her. But I do try and dig out a few things because I feel a bit horrible not giving nothing to her – but her oldest son gets my older boy's clothes.

Tell me a bit about your shoe collection. Do you like shoes?

I do like shoes. My older sister – we're sort of roughly the same size. If she can get a pair of my shoes, she will. I do try and find [her] a pair that, if I'm chucking them out – but she has to wear sensible shoes because she falls over a lot, so she can't wear no heels, but she can't wear very flat shoes – so she needs, not a wedge, but a slight rise shoe and they have to be quite sensible. I don't really have sensible shoes.

But my younger sister has got big feet, I'm afraid.

Describe your favourite shoes.

My favourite shoes are – I have a pair of Jimmy Choos and they're rather high, but they are very comfortable – so... I do love heel shoes because they make you feel taller and slimmer.

You talked about your age. If there was an age box to tick like when you get forms – are you young, middle-aged, older – which bracket would you say you came into, or you feel *that you're in?*

My age bracket – I wouldn't say I'm middle-aged yet. Middle-aged, I'd say, is probably about 50, 55 now – so I don't go into that bracket yet. It's in-between...

I don't know. Young. Not young-young, but... I usually sort of tick the sort of 35 to whatever bracket which, hopefully, I'm in that bracket.

I don't lie about my age.

Myotonic dystrophy is something that can affect all parts of people. It can affect your muscles, eyesight, it can affect the heart. In relation to your own family, if there was a magic wand that could take away one symptom, what would you want it to be?

If I could take a symptom away for my sisters with the muscular [sic] dystrophy, I think it will be to do with the muscles so that they could just be able to walk without their legs hurting them because they can't walk too far because their legs ache. I just think it would be nice if there could be something to help them with their muscles if they hurt.

But they have got something else wrong with them, with their hearts. They've both had to have pacemakers. I don't know which is worse.

Say a wee bit about the pacemakers.

My older sister's had a pacemaker fitted and now my younger sister has had tests and now she's got to be fitted with a pacemaker which is quite sad because you associate them with old people. That's quite hard to think that someone so young has had to have a pacemaker – but if it helps them, and it keeps them here as long as possible, you know, it's good that they find new things to help people live longer and help to relieve their symptoms.

So, if you think about your sisters 10 years ago, and think about them now, have you noticed any differences in them?

I have noticed a massive difference with my sisters in the last 10 years. I think it's a bit longer than 10 years since we first found out that they've got this muscular dystrophy, but since we found out, we had noticed a big deterioration in both of them with their muscles – not being able to walk as they used to be able to; doing simple things takes so much longer.

My older sister – she's lost a lot of her hair. I'm not sure if that's alopecia or through her muscular dystrophy because she's had a bit of a hard life with relationships as well. She's had a bit of a horrible time, but since she's met her husband she's done really well – so she's in a happy place, but it's sad to see, in those 10 years, how much her life has changed through losing her job because of her illness – not being able to do what she used to.

My younger sister – you can see she's deteriorated, but she has to get on with it. Like everyone you have to get on with it, and fight it.

I haven't asked you what you do. You talked about your life. Do you feel the work you're doing on houses counts?

I used to work until I fell pregnant with my oldest son, so I've been a lucky person not to have to work – I've been a fulltime mum. Me and my husband like buying old houses and remodelling them, so that's been my fulltime job because I'm there with the builders, making them sandwiches, tea, running around, building supplies. That's been my job.

And also say a bit about the interiors.

The houses that we've done up are quite large. A lot of people like all the interior and ask if I do interior design, but I don't. I just like looking at places and get good ideas and then I put them into my houses and, yeah, enjoy having nice things in a nice home.

You also referred to it as muscular dystrophy. Do you ever say myotonic dystrophy? Is there anything that frustrates you about it?

No.

Because there are lots of different kinds – but do you say that because people know what muscular dystrophy is? It's more commonly understood.

I just say muscular dystrophy. I've said myotonic dystrophy before, but I just presumed it's the same thing. I just say 'myotonic' and people go, 'What's that?' I go, 'Oh, I think it's muscular dystrophy'. Most people have heard of muscular dystrophy because it's getting more well known. Ten years ago not anybody really heard of that – but, yeah, I have to explain to friends that I got two sisters with muscular dystrophy. I say 'muscular dystrophy'. I don't know if it is myotonic or muscular dystrophy – but, yes.

Because my younger sister's got a disabled child, and everybody is on Facebook, and they see pictures if I've commented and they go, 'Oh, who's that?' and I go, 'It's my niece, and she's got muscular dystrophy'. They go, 'What's that?' and you just go, 'It's like a muscle-wasting thing' and they go, 'Oh'.

And they don't ask anything else?

No. They just go, 'Oh'

So you don't find yourself having to explain it?

No.

Sally and Lucie

Sally, 32

Age at diagnosis: 29
Carlisle, 1 March 2015

Jackie: *If you could start by saying your name, how old you are, and where you're from.*

Sally: My name is Sally and I am 32 years old. I'm from Carlisle and I live in Carlisle.

Thank you. Do you have any siblings?

Yes, I have one sibling – an older sister. She is two and a half years older than me and her name is Lucie and she lives in Lincoln.

Me and my sister – we got on very well as kids. We were very close when we were growing up because she was the only sibling I had – but as time has gone on we have grown apart a bit, but we still do talk. We don't look much alike. We look very different because I have myotonic dystrophy and she doesn't.

I was very jealous of her when I was growing up because there were a lot of things she could do and I couldn't do, and she was very clever academically, and I wasn't – so, I was very jealous of her growing up because she could do a lot of things that I actually couldn't do because of the condition.

Were you aware of your condition when you were at school, when you were younger?

No, when I grew up, I wasn't aware that I had a condition called myotonic dystrophy. When you have things that happen to you – like, if you are born with asthma, or you're born with eczema you just kind of assume that's who you are, or if you need glasses or hearing aids you just accept – that's who you are. So, a lot of things I just put it down to just 'me'. That's the way I am; the way I was born – so I didn't actually realize that I had it, but it actually does explain a lot of things now when I think back to all the symptoms you get from this condition, it does explain a lot of the ones that I do have.

That sounds very much like me and my sister. If you could talk about some of those incidents that joined up or made more sense once you got your diagnosis?

When I was told I had myotonic dystrophy, I was told to have a look on the Internet about what symptoms I had, and talked to specialists about it, and when I looked at the symptoms, one of them was, when I was a little girl, I had speech therapy and for a lot of kids, that is a symptom.

When I was about 11 I had an malocclusion in my mouth and the dentist had to have a look and said, 'Well, either it will get worse, or the occlusion will get better'. I actually couldn't bite at the front. If I had to bite, I used to have to bite at the side of my mouth because I couldn't bite at the front. The doctor said, 'What we can do is, we can do two things: we can either realign your teeth one way and you have an operation where we'll break your jaw and make your teeth meet, or we can realign it in a different way. We don't break your jaw; your teeth won't meet, but they'll be close', basically – so I had to decide myself when to do that. He said 'We'll have to wait until you've stopped growing', because things change as you grow. So, I decided I wanted to have the operation when I was 17. He said to me, this is what's going to happen through the operation – blah, blah, blah. When I went to sleep on the Sunday night in hospital, that's the last thing I remember before waking up on Thursday, the following week.

So I had the operation on the Monday morning, but I don't remember being woken up on Monday morning; I don't remember going down to theatre – I don't remember any of that – and what happened was, I was asleep for so long because I had a bad reaction to the general anaesthetic, and that's also another symptom of this condition. You can be sensitive to general anaesthetics. The doctors didn't know what was going on and they said, 'Well, she'll either come through or she won't'. I think it was kind of 50:50, really, they thought.

The anaesthetist had probably assumed everything would be OK. He never came back to see me, which I think is bad for a doctor. He should have tried to figure out why I had such a bad reaction to the general anaesthetic because the majority of the public don't.

What happened was, because I couldn't wake up enough after the operation, I couldn't cough out all the blood that was in my mouth so, instead, I inhaled that back down into my lungs and, as a result of that, my left lung collapsed, so I was on a ventilator for a few days.

Then I managed to come round on the Thursday afternoon, it was, and I remember waking up and thinking, 'I have a tube down my throat'. I thought, 'I don't remember this being explained to me and why is it Thursday? I went down on a Monday'. So, the doctors came round and said, 'Well, we're not sure if you're going to be able to breathe by yourself, but we'll take it out and see what happens'. And I did. I was able to breathe by myself.

They just did not know what happened. They said it was 'one of those things', and that was that – that was the end of that.

So that occlusion, malocclusion, in my mouth is a symptom, and the operation where I had a bad reaction is a symptom of it.

And then, when I was a teenager and developing and when we get cramps from periods and things like that, and I used to get bad stomachs – not just that time of the month – any time of the month – any time it could happen. I used to get really bad stomach pains and diarrhoea quite a bit – so I just kind of thought at one point, 'Oh, I think I might have IBS' – but I never got it confirmed, I just thought to myself that's what it is – so I just take Diclofenac for the pain and then if I have diarrhoea at the same time, I just take Imodium to get rid of that – but I've learned that IBS, that's a symptom. You can get really bad stomach pains with it – so that's another symptom of it.

I was always, what you would call, 'lazy' as a kid. I just had no get-up-and-go. I like doing sport and things like that, but I didn't do the Duke of Edinburgh [youth achievement award]. I just couldn't be bothered to go walking. I hate walking. I tried many times as a kid to get out and go for a walk. I just didn't want to do it.

I used to get very tired and my parents just thought, 'Oh, she's just being lazy', and I just thought to myself, 'Why am I lazy, though, because none of my family are like that?' And I just wondered for years why I was so tired – but it's just another symptom – this apathy that you get, and no get-up-and-go. That's another symptom of this condition.

Also, the heavy eyes is another symptom but people just thought I looked like my dad with that because we found out that my dad has this condition. That's how I got it. My dad has heavy eyes, sunken temples, and my dad had a droopy eyelid for years and it got worse and worse as the years went on, so eventually, recently, he had it lifted. It's pretty similar to the other one, not quite identical, but pretty good. Yes, so that's how we found out. After all these years things actually make a lot of sense now.

They also said at school that I had learning difficulties. They said, 'You're not as bad as dyslexic people, but you do have a difficulty'. They said it is a visual/spatial difficulty where, if you asked me to draw, say, a chair, I could try, but it would not look like a chair and if you didn't tell me what to draw, you probably would have no idea what it was that I was trying to draw. I was always really bad at art. I used to get really bad marks at art – so that's another symptom that I've got.

I've not been very clever at school. I wasn't very good at school. I didn't learn a lot. So, I was very bad at school.

Also, I have difficulties sometimes understanding instructions. If people tell me where to go – like, say I'm going somewhere for an interview or a friend's house if they've moved, I always go a few days before and take someone with me so that I know where I'm going because, even if I have been before to a place, sometimes I still get a bit lost. I sort of can't quite get there.

When I found out about my condition – I was 29 when I found out in 2012. The reason I found out about this condition is because I had a baby and he was born at 27 weeks and four days, and so I knew that, him being that early, he would need help to start with breathing and stuff but then, after a little while – after a week or so – the doctor said he isn't needing much help breathing, but he won't actually make an effort when we take him off the ventilator to breathe; and he's not moving around very much. So they said, you know, 'We don't know what's going on here'.

When I had the baby I was in Lincoln Hospital at the time. I was actually away on holiday visiting my sister.

He was born early at 27 weeks and four days' gestation (and premature labour is another symptom of this condition), and the doctors... I knew, him being born that early, he might have difficulty breathing and a lot of thing because of being born early – so, the doctors said he is OK, but they said he isn't moving around a lot; he's not making any effort to breathe by himself. 'I'm going to take him off the ventilator' – so they weren't sure what was going on.

I had the baby in Lincoln. They could save the life of a baby so far, but they can only do 32 weeks plus – so they transferred him to Nottingham, then a day later I was transferred to Nottingham Hospital because I still had to have injections because I had a C-section and they said, 'Your baby is OK, but he is a bit poorly'. They said, 'He's got two kidneys, but one is bigger than the other'. I think the big one compensated the smaller one and he had a hole in his heart, but they said, 'You know, he is doing OK apart from that'. Then they decided to transfer the baby to Newcastle Hospital which is closer to where I live with my family – so they said 'We'll transfer him to Newcastle Hospital and there is a place on the grounds where you can stay while your baby is in this family unit'.

So I went there and I spoke to the doctors and one doctor shook my hand and said, 'Hi, I'm Dr So and So', and then he realized something. He didn't tell me at the time, but he said to me, he said, 'When you shook my hand, you didn't let go of my hand very quickly'. My hand kind of locked when I tried to let go and my hand just very slowly released – and he'd had experience of this condition before with some babies, so he thought, 'Right, the baby is not moving around; not trying to breathe; she didn't let go of my hand very quickly', and he said, 'Your facial expression is typical of someone who's got it' – so, what they did, before telling me anything about it, they thought they'll do a blood test on the baby to test that specific gene number and see if he's got it and they did, and they said, 'Yeah, he's got it' – so they told me all about it and that he got it off one of either me or my husband, but they said, 'We think it's you', because they've had previous experience of it.

They said, 'Your family is going to have to be tested as well to find out who's got it, and who hasn't because it can go through several members of your family' – so, they tested me and, yes, I had it.

Then they told [me] I had to let the baby go because they said he won't survive no matter what. They said if you don't take him off the ventilator, he'll die on it. So we had to let him go at four weeks old – so we did that and it was a very sad time, but it was better for him because he would have had a not very good life, I guess, because he would have had a lot more symptoms because the normal way of this condition is, you get worse as generations go on, you tend to get more symptoms so it is a lot worse and, apparently (I've been told this) it's worse if you get it off your mum than if you get it off your dad, for some reason.

It turned out my dad had it, not my mum. The nurse that came to take blood said, 'I can tell who's got it, but I have to take both blood samples to make sure'. So, my dad had it and a lot of people used to say I look like my dad, but I never thought that. I thought I looked more like my mum because I've got Mum's eyes, Mum's teeth, and Mum's sort of figure, but my dad... My sister looks more like my dad...

But from about here [*indicating eye level*] upwards I look like my dad. I've got the bags under here; the sunken temples and sometimes I get a look on my face and my mum says, 'You look like your dad when you're pulling that sort of face'. So, I do have a bit of my dad in me.

My dad – the symptoms with him... I used to go over on my right ankle when I was walking through town. I kept thinking, 'Oh, I've just got weak ankles' – that's a symptom, and my dad always falls over his own feet, you know. My mum just thought his feet are too big because he used to fall over his own feet; and my dad had a droopy eyelid for years; and my dad is very, very thin – you know, stick thin. His waist size hasn't changed since he left school, and my dad is almost 70 – so that's another symptom. You can be really thin, or you can be really obese as well.

When my dad got tested and found out he had it, he had two sisters and they both had to be tested but one unfortunately passed away before we found out about this – so, his other sister got tested and, yes, she has it, but we don't think the other one did just because my auntie looks like a double of my dad, just a female version. She's very thin, really heavy eyes, she also looks very tired – my Auntie Jane. She always looks tired.

So, when she was tested and found out she had it, her kids had to be tested. She had two girls and a boy. The two girls have got it, but the boy doesn't. Now, my cousin (who is my age) – she was told that she had Möbius syndrome[1] when she was born (I'm not exactly sure how you say it) because she was born with bilateral hips where her feet were like that [*holding hands horizontally with fingers pointing inwards*]. She's had them altered now, so her feet are straight but that's a lot of operations she's gone through. She's got muscles in her face, but she can't move them – so, yes, she's got no expression on her face at all. It's just a normal expression.

The other girl that's got it – you really can't tell how much she's got it, really. She's not that bad, but I don't know about things that's gone on in her life that... Yeah, so my cousin was always... Obviously, there's something wrong with her, but we just thought it was that but, actually, it turns out it wasn't Möbius syndrome, it was myotonic dystrophy that she has. That was all the siblings done.

My dad, we had to figure out whether it was his mum or his dad that had it and they had both passed away but we think it was Dad's mum that had it, and she had two sisters, and her sisters are fine. They haven't got it, but we think maybe my gran did because she used to find it sometimes hard to get up in the morning; she used to get quite tired herself – but as you get older, you just kind of think it's old age when you start to get tired, but we think my gran possibly had it because she was stick thin; she looked like my dad.

And my cousin who was really bad, she is really thin.

1. Möbius syndrome is a congenital neurological disorder which is characterized by facial paralysis and the inability to move the eyes from side to side.

So that's how we found out about it, really, just from my baby. I really should have found out when I was 17 when I had a bad reaction to the anaesthetic, but the doctor did not come back to figure out why and I know, for a fact, that anaesthetists have heard of this condition because they need to because if you need a general anaesthetic for any operation, they need to be careful how they give you a general anaesthetic in case that happens again – you have a bad reaction.

My cousin was misdiagnosed for over 20 years as well with her condition – so that's how we found out about it – and unfortunately we had to find out the hard way with my baby passing away.

So, you now have a new baby. If you can say a little bit about your son?

Well, after my baby passed away, my husband and I decided we really wanted a child – so we decided, maybe we should try again, but my doctors did tell me that if you try naturally to have a baby, you may have to have a termination. They said, 'We can do a special test when you're about 11 weeks gone where they stick a needle into your uterus and take some cells to test the baby', and they said, 'If the baby has myotonic dystrophy, the best idea is to get rid of it because the baby won't survive and it will be a whole lot of heartache for you again – so you have to get used to the idea, can you cope with having a termination or not?' They said, 'If you can't, there is another thing you can do which is where they take the eggs out, put the good eggs back, and then see – but that doesn't always work and it can be costly – so it depends what you feel like and if you really want kids and it will make your life better or whatever...'

So me and my husband decided to try and see what happens, and I get very sick about six weeks into the pregnancy – so I knew I was pregnant because I got that sick feeling again. My husband and I had to kind of calm down and not get too excited and didn't tell very many people – just my mum and my dad. I didn't tell my sister or anybody else – just in case the baby was not OK.

So we had to wait a few weeks, then when I was 11 weeks I went to the RVI Hospital in Newcastle and they did special tests and the baby looked fine compared to the first one when I saw the screen because when I had the first baby, I had to go for a special test after 12 weeks because they said there was too much fluid behind the baby's neck – so all these things were telling me there was something wrong with him and they also told me later on that he had a ventriculomegaly which was, part of his brain was too thick, you know – it was thicker than it should be – so, all these things were pointing to there being something wrong with him, but they couldn't tell me what they may be able to do or how severe or otherwise it was. I decided to have the baby because there was a 25 percent chance that there was something wrong and 75 percent he's fine – so, that's why we went ahead with the first pregnancy.

Anyway, so I looked at the screen and I said, 'What about the fluid behind his neck?' The doctor said, 'No, it's fine. It looks OK, but that doesn't mean anything, really'. So she said, 'We'll do the test'. So I had to have the needle stuck into my uterus and I had to wait a couple of days just for the initial test of myotonic dystrophy and Down's syndrome and she rang me on the Thursday and said, 'The baby is fine' – so that was brilliant – and she said, 'We're going to test every chromosome to make sure there is nothing else wrong at all'. So they tested every chromosome in his body. They said, 'There's nothing we can tell from the chromosomes. He is fine'. And they said, 'Do you want to know the sex?', and I said, 'Yes', and they said, 'Well, it's a boy' (because my husband was always wanting to know what it was – he was dying to know). Well, it was fine. It was brilliant, because then we could start making plans and stuff.

The pregnancy was going perfectly fine and everything was good after I stopped being sick and then, when I was 24 weeks gone, I was just in bed one night and it was about 3 o'clock in the morning, and I felt a trickle down my leg. I thought, 'Oh, gosh! Am I weeing myself in bed?' The light was still on at the bedside so I just pulled back the covers and I was sitting in a pool of blood and I was like, 'Oh, my goodness!'

So, after we rang the hospital they came to get me. The hospital where I live in Carlisle, they couldn't deal with anything like that at that stage, so they said, 'Right, the nearest hospital we can send you to is Sunderland. They've got a bed for you and the baby if the baby decides it wants to come now'. So I had to be sent to Sunderland Hospital.

I was there for six days and the doctor said, 'We'd send you home, but you don't live anywhere near us here'. So I said, 'Well, I'd rather deliver the baby in Newcastle anyway' because they found I had a placenta praevia where the placenta comes into the cervix entrance so the baby couldn't come out naturally anyway and they said, 'You're going to have to have a C-section and it's better to have an elective one than an emergency' – so they said, if you can find a place to rent in Newcastle until the baby is born because, they said 'You've had a bleed; the likelihood is you'll have another'.

We found a place to rent in Newcastle and I was there for just about a week, and then I had another bleed when I was 30 weeks – so they said, 'We're going to take you into hospital and keep you in hospital. I know you're not far away now – but it is just to be safe to keep you in'. So I was in hospital from 30 weeks until I gave birth to him at 37 weeks – so that was quite a long stay in hospital.

He decided to come a week earlier than they planned, but it wasn't such a bad thing, really. At least I was there. They monitored me for an hour in case it was not a proper labour – just Braxton Hicks – but after an hour, they said, 'We'll get him out because he is 37 weeks; he is viable enough to live; he should be fine when he comes out. We can't see any problem with it'.

So I went in for another C-section and this time they gave me the spinal injection and after a few minutes it hadn't worked. I would still feel the spray. They kept spraying me with this cold spray. 'Can you feel that?' 'Yes.' So I had to have another spinal injection, and then that worked and then they got the baby out and he was fine. I knew he was fine because as soon as he came out, he started crying and then I cried because I never heard my first baby cry for obvious reasons. He was too early to actually do anything like that.

So, yes – we decided to try for a baby and we got lucky, really. It was a 50:50 chance as to whether or not he'd be fine – so I'm really happy with it.

Can you say lastly about what his name is, talk about his christening coming up?

My baby – when he was born at 37 weeks, he was 7 pounds and 5 ounces. We had a day or so to think about his name because we had a few names we had chosen and actually, I think he looks like a Kane – so my husband and I both like the name 'Kane' – so I thought, 'Great – that's it: "Kane"'. That's spelt K–a–n–e. And then we thought about a middle name because I thought it is nice to have a middle name if you don't like your first name – so we decided to think about the initials of his name because, sometimes if you say initials, it makes up a word (like my cousin, since she got married, her initials are S A D – 'sad'). The surname begins with a 'C'. So we thought what initial would look good, and we thought the initial 'D' would look good – so then we had to think about different names – Daniel, David... We actually ended up with the name 'Dylan' – so it's Kane Dylan.

He is just under six months old now – he'll be six months in a few weeks and his christening is coming up on the 15th of March and he's actually getting christened at the church where my husband and I got married in 2011 – so that will be really nice, and most of the family will be able to make it which is nice – so we're looking forward to that.

Lovely. It will be a good day for everyone. I want to go back to talk about your sister again. What I'd like you to do is... I've never met your sister. I'm going to meet her. I'd like you to describe her for me. So, if you could describe what she looks like and also, what she does, just a bit about her.

OK. Well, my sister – she is two and a half years older than me. I'm about half an inch, maybe to an inch, taller than her, but not much. Well, my sister doesn't look much like me at all. My sister is very thin. She's basically got the figure of a supermodel where it's really abnormally thin – but that's just the way she's born, really. So, she's abnormally thin; she's very pretty; she's got roughly the same colour hair as me – maybe a bit lighter or a bit darker because she does tend to dye it a bit.

She is very clever as well. She is actually a barrister – that's her job – so she works in London, but she lives in the countryside. Her husband is a doctor. He didn't want to work in a big city hospital because he doesn't like the way they are, so he wanted to move to a small hospital. He's actually an anaesthetist and he works in the intensive care unit in Lincoln Hospital – so that's why they moved there because they thought about where they would like to live – so my sister lives there.

She is very intelligent and always has been. She's always been very bright academically; always top of her class; and always could do a lot of things that I couldn't do – which made me very jealous of her – just certain things. I was like, 'Why can't I do that?'

Like I say, she's very intelligent and she got married the same year I got married but, like I say, we don't really look alike. You'll probably have difficulty picturing us as sisters.

We do get on well together. We are close. She is always there for me; she always helps me if I need anything.

My sister is very independent. She doesn't really contact you unless she wants to know something or something's coming up. She's very independent and doesn't tell us a lot of what is going on – nothing seems to bother her – faze her. She is very just... 'Oh, I'll do this; I'll do that'.

My sister used to be very stressed around exam times at school whereas I was just laid back because I knew I wouldn't do very well – but she thought, 'People expect a high level from me', whereas me, I'm very laid back. I'm actually too laid back for my own good at times. You know, nothing fazes me; nothing bothers me.

Yes, so my sister – I was very jealous of her because, for example, for jobs as well. I used to find some jobs... They say, 'Why can't you do this; why can't you do that?' and, 'You're not working very fast today'. But, obviously, they don't know about my condition and there's nothing I can do about it.

I've heard of people with this condition who do have difficulty in keeping a job just because they're not quite as fast and their muscles can't work as fast as other people's – so they think, 'Oh, they're not trying hard; they're not good at this job', and let them go.

My sister has always been very good and always helped me out. If I ever needed help with anything, she was always there because I was very shy as a little girl, but she was more outgoing and boisterous. My sister used to [talk to] me when I was a little girl because I was always with my sister and she was always doing things for me – helping me out with schoolwork and just anything she can figure out how to help me with, really.

Like I say, she was a very good sister to me – helping me out with things and so, like I say, she's very lucky. She doesn't find anything hard; she's very bright – yeah, she's found life really easy, and because she knows that she doesn't have myotonic dystrophy, she's 'Oh, well, if I have a kid, that will be fine. Nothing is going to happen to my baby'.

Yes, she understands a lot more now as well – just thinking, 'Oh, yeah – that's why Sally's like that' – because of the condition.

She's very good to me. She is always there for me if I need her. I can go and see her. Yes, we are kind of close. I have to be, really. She's the only sibling I've got, you know.

But like I say, she doesn't look anything like me and she's very thin and she's very pretty – so... Yes, I did used to be jealous just because of the way she was, but now I realize, these things happen. You can't change them.

Could you say about what her specialism is? She's a barrister?

Yes, she's a barrister. When she was choosing her options when she was a bit younger in school about what she wanted to do, she did want to do law, but the thing that put her off was the first two years was just case studies and that would kind of bore her because they're just very basic things – so my sister thought, 'Maybe I should do something else instead', and someone told her about a conversion you can do where you do the first three years in one year.

So she decided to do a degree first because if you have to have another degree, it helps you getting a better job. So my sister said, 'Well, what am I really good at?' She was very good at chemistry and physics, so she thought, 'Well, I'll do a chemistry degree' – so, she did a chemistry degree for three or four years and because she's very bright, she obviously got a first [a first-class honours university degree] in it. She was very happy when she got the first in it and my parents were telling me, 'Oh, it's wonderful. She's got a first'. And I was like, 'And you're surprised – why?' It was kind of obvious she was going to do very well.

So, she did the chemistry degree and then she said, 'I'll have a gap year'. When she did the gap year she just did a bit of temping, a bit of reception work... She actually managed to get a temping position in a doctor's surgery. She did a bit of travelling – she went to Taiwan and Ghana. She worked in an orphanage for a little bit and then she came back and said she wanted to do the conversion to law – so, she did two years of law and after she passed – got called to the Bar and passed her exams – she had to find a pupillage in one of the law firms and she actually got quite a few offers, which was good. So she decided the one she wanted to work for, and then she has to work for that company for a year and then, at the end of the year, they'll decide if they'll take her on or not or take some other pupil that had pupillage there – take them all on; or take one or two on – and then they decided to keep my sister on – so she did really well.

She is self-employed. She works for a company that specializes in construction technology, but she is self-employed, so she only gets paid for the work she does – but yes, she loves it. She likes the hard work and there's a lot of travelling with it. She actually had a case to go to in Dubai – so she spent quite a few months in Dubai getting evidence for this case which never actually happened in the end.

Yes, she loves it; it's hard work – she works hard most days, five days a week. She works late at night. She works from, like, 8 in the morning until 8 in the night sometimes, or even later. She doesn't have to go down to the office every day in London because she lives in Lincoln. It's only an hour and five on the train – commute-wise – but she only has to go down if she has a meeting or needs some information, but all the work she can do from home. She's got an office at home where she has her computer and everything around her. Yeah, she really enjoys it. It's hard work, but she really does enjoy the job she does and she gets paid really well for it too.

You talked about when you were at school and how you didn't feel you necessarily got on that well while you were there. Did you go to university after? Just say what you did, and what you do now.

After my GCSEs [exams] at school, I decided to leave school then because I thought, 'Well, I didn't do very well'. So I thought, 'What am I going to do now?' I thought I knew what I wanted to do. I thought I wanted to be a veterinary nurse. I thought... I looked up courses and there was a course at one college that I couldn't start until I was a bit older, so I thought, 'What will I do for a year?' So, my mum and dad always helped me to decide what to do, so Mum said, 'Well, you're not 100 percent sure you're going to like it when you get there'. So Mum said, 'Why don't you do an animal course instead and see what you think?' So, I thought about going to a college that's very close to here. It's only about 30 miles away – so I can still live at home because I'm very much a home person. I didn't want to leave home at the age of 16. I wanted to stay at home, whereas my sister couldn't wait to get away from home. She left as early as she could. She was 18 when she left and went to the University of London.

So, after my GCSEs – because I found them very hard, I thought, 'Right, I want to go to college'. So I went for an interview at college and they said, 'It doesn't matter what you get in your GCSEs. You can still get all the lowest grades and you can still do a course here. It doesn't matter what you get.' They said, 'Depending on what you get, it depends which course you can do. There is a three-year animal course, a one-year or a two-year course'.

I managed to get two Cs in my GCSEs so that meant I could do this one-year course where you get a National Certificate in small animal care and then they said, 'If you pass that, you can do the second year of the diploma'. Basically, the only difference was, in the Certificate, you do a lot more practical work in the first year – so if you go on to the second year... And people who were doing their diploma in the first year – you don't spend a lot of time with them because they do a lot more practical work in the second year.

So, I did the Certificate and I passed and then I went on to do the Diploma because, on the first year, you had one day at college where you had to work in some animal capacity – whether it be a kennel, a cattery... That way you get practice in all sorts of animal places and I managed to get a placement (because they didn't have to pay me – it was just for free) where I worked one day in a veterinary practice.

At the end of the first year I realized, it was good – I enjoyed it, but I realized that's not what I wanted to do. I thought, 'No – that's not what I want to do'. My parents were right that I wasn't actually 100 percent sure.

So, after I passed the certificate, I thought, 'Well, I'll do the second year of the Diploma to get the National Diploma and just see what other options are available'. There was the RSPCA [Royal Society for the Prevention of Cruelty to Animals] – there are different things that are available to you.

The RSPCA was an option, but you had to be 21 to start working there, and I would not be 21 when I've finished the course. I actually did quite well in the course because a lot of the marks were based on coursework, and I found coursework easier than exams because you don't have to remember things. Coursework is in front of you. Yes, you get these books and whatever else and if you do well, you get the marks for it. So there wasn't much exam work to do for it, so I did really well.

So, I did the animal course and I realized after the second year, 'Maybe I don't want to work with animals; maybe I just love animals'. So I thought, 'Right. What am I going to do?' So I had my summer off when I left college and I thought, 'What am I going to do now?' I wasn't sure – so, I was actually going to sign on [apply for unemployment benefit] to get some money but the Job Centre said I had too many savings, so I couldn't get any money back to get my stamp [National Insurance contribution] paid.

I was only doing that for a few weeks, looking for a job, and this training company got in contact with me and they said, 'We're a training company where we can let you do NVQs [National Vocational Qualifications] in some sort of business like either [ ] administration, retail, hairdressing, teacher assistants – whatever. They said, 'If you come for an assessment, we'll let you know whether you can do Level 2 or Level 3 depending on how good you are in the assessment'. The man that took it said to me, 'Well, you're very good'. He said, 'You're only off Level 3 on a few of them, but unfortunately, because you are just below with a few marks, you have to do Level 2 first. You can do Level 3 later if you want'. So he said, 'You can do Level 2. What would you like to do?' So I said, 'Business administration. I'll try that'. He said, 'What you do, you find a placement in a place. They don't have to give you a job. We give you a training allowance every week or every month depending on how often you want to, really'.

So I had to find a place where they wanted to offer me some work and they got me free, basically, so they didn't have to pay me at all. So I found a place at an insurance brokers and I worked there. It was commercial insurance that they did. It's big places and they had a few [fleets] that they looked after as well and I really enjoyed it – not necessarily where I worked, but the actual way that an office works – so that was really good. I had a really well-rounded NVQ as well because there wasn't any exams – just a few pieces of coursework, really, to gather all this evidence.

I really enjoyed that and at the end of the 18 months of doing that NVQ Level 2, they said, 'Unfortunately we can't offer you a job. We're fully staffed'. So that was fine. I just had to find another place to work then.

I had a few jobs, and then I went to another place and they wanted me to do a Level 3 while I was there, so I did that. You get better jobs if you have Level 3 and then, after a couple of years there, there was no way in that company that I could progress any further. I was as high as I could get, so I thought, 'Well, I'll find another job where I can actually progress and get higher up'.

The best job that I found was that I managed to get a job at the Department of Work and Pensions in Carlisle and I loved the work I did there. I loved the job I had, I loved the people I worked with and I was really happy there. The only reason that job ended is because I got made redundant because the office closed. Everybody there got made redundant.

That's how my last job ended. I was actually pregnant with my first son when we found out the office was going to close – so I did not actually end up going back because, by the time I was due to go back, the office had closed – so that was the end of that, really.

So, yes, I didn't want to go to university afterwards. My parents couldn't afford it at the time to send me because my parents were busy buying a new house at the time but I would have gone to uni after college if I wanted to, but I didn't want to, so it was no big pain about it, really.

I didn't want to leave home and I needed a lot of help when I do do academic work because, like I say, I'm not very bright with academic work. My parents always have helped me a lot. I just thought, 'Well, it's just one of those things. I'm not very bright', whereas my sister never needed any help whatsoever.

Yes, so that's what I've done. So I've been in business administration since I left college, but at the moment I don't have a job because I've just recently had a baby anyway – so...

You're a mum.

Yes – a full-time job.

I'm going to ask you a few questions about ageing. When you think about those box categories that you sometimes get on forms, how would you describe yourself? Would you think you were young, middle-aged – you know, those kinds of categories.

I think I'm a young person because I'm only 32 – so I'm not middle-aged yet or anything like that – so, I just class myself as not really young, but youngish; not as a teenager or a 20-year-old, but just a little bit older than that.

You've been through a lot in the last five years. When you look back, do you feel you have got older, or do you feel the same as you did when you were a teenager?

When I was a teenager, I think I was a very young teenager, whatever age I was – 16, 17... I think I was always on the very young side; I wasn't very mature – but I really enjoyed my time as a teenager, as you do – going out and everything – all things compared to that. I always thought I was young. I always felt very young, and I really don't feel too different now. I've matured a lot. You do grow up and realize things aren't always as easy as you thought they were going to be and things happen to you and you have to deal with these things. You do. You sometimes have a hard time.

I think when you're a teenager, life is always going to be great, but it doesn't always work out that way. You do get hard times for whatever reason. You get depressed or have a hard time at a job or you lose a child – anything. You can't really prepare yourself for the bad times. You just think life is going to be great.

I still feel quite young in myself. I don't feel like I'm elderly at all. I just don't feel old. What I used to think when I was younger, I thought, 'Oh, 30! Gosh, I'm not looking forward to that age! That's really old' – but when you get there, you don't actually feel that older. I don't really feel that I've left my twenties yet. I still feel quite young.

But like I say, I've learned a lot since... I've moved out of home; I had a job; I've had to support myself and things – so, I have matured a lot.

Yes, I still feel pretty young in myself even though, in the last few years, my condition has got a lot worse. A lot of things I never used to find hard, I do now – for example, I've always had a problem with buttons sometimes, especially with new clothes. Button holes are sometimes very stiff, and I find it very hard sometimes to do buttons up or open bottle tops, and carrying the baby around I sometimes... I tend to sit down with him and let my husband do more the heavy lifting work.

Some things I just cannot lift because they're too heavy for me, and I've been told to be careful what I do because some things can make the condition worse – like physical exercise. The doctors told me, 'Pilates is good, and swimming, but going to a gym won't necessarily help you'.

Yeah, over the last few years, my walking has got a lot worse. I can't walk very far without getting breathless. I can walk a few hundred metres and that's me done. My legs start to hurt, my back starts to hurt – so, a lot of things have got a lot worse for me.

I was told that, you know, 'As you get older, your condition might deteriorate because it is progressive'. They said, 'You might deteriorate, you might not. Nobody knows. Everyone is different'.

But like I say, since having the baby, things have got a bit worse for me – but I have my husband here to help me; my parents are close by; his family are close by, and they all know what I've got although not many people have heard of it before. I've never heard of it before, and my dad who is a GP and a medical director (or was, before he retired) – and he never heard of it – so, it's just the awareness of this condition, really.

Most people have never heard of it even though it is not as rare as some conditions. For example, Huntington's is rarer – but a lot of people have heard of Huntington's disease because it's been publicized, say, on a soap or something – whereas this condition, they can't really publicize it because there are so many different symptoms you can have.

They can't say, 'These are symptoms of it'. They can't say that because everybody has different symptoms – so they can't say, 'Blah blah, they said on TV' – but that symptom may not happen – you don't know because there are loads of symptoms in this condition so like I say, it really is hard to publicize it.

Like I say, it does explain a lot of things when you think about just my life. It does.

Good. So that is the ageing thing. My sister and I used to always laugh… She was always determined to never have any grey hair. My grey hair has come in. I'm younger and I am not allowed to have grey hair – these things that sisters do. Today, there are so many different treatments available for women to prevent ageing. What are your thoughts on ageing, particularly to do with your appearance and things that you maybe consider doing and things that you've seen other people do and thought, 'I wouldn't do that'?

What I think about ageing is that I know myself that I look really young for my age. I've always been like that. I still get ID'd, say, when I have a night out, I have to take my driving licence with me because I always get asked my ID. People think I'm a lot younger than I am but when I tell people how old I am, they never believe me. They say, 'No, you're way too young'. Yeah, I always look younger than I am.

My sister – she doesn't look older than she is, but she does look her age, really, whereas I've always looked younger.

I'm not too worried about getting older because my mum, she has a very young appearance as well – so I think that's probably where I get my appearance off, it's my mum.

I think I look really young compared to my husband, because my husband and I are the same age but we were on holiday a few years ago, in Egypt, and somebody said, 'Oh, you're here with your dad', and I was just like, 'No! No, I'm not!'

Yes, so I'm not worried about getting older, really. I'm a bit worried that my condition could get worse and I'm going to need a lot more help. I'm a bit worried about that side, but I'm a very positive and optimistic person. I don't worry about anything until I have to.

And as for appearance, I don't use anything out of the ordinary. I use Oil of Olay for my skin after I've had a wash. I just put Oil of Olay on my face, but I don't actually use much. I put a bit of make-up on so I look OK – but like, for skincare products, I don't go mad like some people do or anything, and I don't think I'll have any operations in the future to make my appearance better – like have a facelift or make the cheeks look bigger or anything like that. I don't think I'll do anything like that because of my problem with anaesthetics anyway.

I'm not in the public eye, so I don't really need to care about my appearance as much as some people do. I would never think of having anything done unless it was a real bad cosmetic problem that I saw myself. I put make-up on and things like that – but as far as going for cosmetic procedures, I'm not interested at all in them.

So I need to talk about shoes – it's quite important. What kind of shoes do you like, and have you had to make any adjustments to your shoes recently? Things like the sort of things that you're able to wear.

When it comes to shoes – when I was younger, I used to like shoes with a little bit of a heel on – as you do – you like to have a bit of a heel on, and I used to like boots – ankle boots and knee-high boots – when I was going out and some of the heels were bigger than others and it was alright. I used to find them alright. I had no trouble with heels on, but when I was working, I'd always wear flat shoes because, if you're on your feet all day, and you have heels, I know your feet can get kind of sore because I've done that before. It's not a good idea.

Now I do find it really difficult to walk in heels, so I wear a lot more flat shoes or trainers because heels – I'm more likely to break my ankle if I wear heels these days. I can wear things with a little heel, but I tend not to. I tend to have a lot more flat shoes than I ever did. I never thought in my life I would ever wear flat shoes. 'No, that's an old lady's thing' but now I realize I do have to wear flat shoes because it's safer for me if I wear flat shoes because I also do have ankle supports for my ankles now. I wear them when, if I go out for a walk or something, or if I'm going somewhere because it makes me feel a lot more steady on my feet. If I'm walking without them, you can guarantee I'm going to go over on one of my ankles. It's normally my right ankle I'd go over on. I never fall over, I just sort of stumble.

In regards to shoes, I'm not a shoe fetish person – but I do like a lot of boots, I must admit. I do have a thing for boots.

But like I say, I have to go for a lot more flat shoes, whereas my sister, she has a lot more shoes than I'll ever have and she doesn't wear big heels. She wears little heels, but she can get away with big heels if she needs to but I don't see the point. People who wear killer heels – I think they're mad because they're going to cause themselves some serious damage when they're older because you're not born to stand on tip-toes – so, people that do do it, I think they're just going to have a lot of problems and I don't need any more problems with my body than I've already got.

So just the last two questions. You've got this young son and your life's changed a lot recently – what are your thoughts on the future and, I guess, something that you're excited about, and anything that you worry about?

For my future, I'm not too worried about it. Me and my husband, we have been together 10 years now. We've only been married for three, coming up four years – but we have been together a long time and like I said, we didn't know about this condition when he met me. He's still around. He hasn't gone because something new has come along – so we're quite happy and he knows – he understands. He helps me a lot and we're going to teach our son when he gets older about this, saying he may need to help Mummy with some things. I don't really want to put too much pressure on him so that he has to stay here and look after me. I don't want him to do that; I want him to have a life – but, you know, just so that he is aware that, sometimes, he may need to stay back and help his mum with things if Dad is not around – but like I say, my husband also doesn't work. He's been signed off now for a while off work – so he's obviously here to help me.

If he can't be here one day, for whatever reason, and I'm by myself and the baby, my parents have agreed to come down and help, or a lot of my husband's family live around here – so they can always help out as well.

So, as for the future, I'm not too worried. I am a bit worried that this condition can get worse; I'm worried about what might happen because I know that you can get cataracts early on with this.

I have first-degree heart block because when I have an ECG every year, they tell me I've got a first-degree heart block. It's not a major problem, but... You know, it's just one of those things you can get with it.

So, I'm not too worried. I am a bit worried about illnesses, though. When I get ill, I get worried if it turns out to be worse because they say, 'If you get a chest infection, you can end up in hospital with it', and six weeks after he was born I thought I had a chest infection. I got sent to hospital. It turned out I had pneumonia, so I was in hospital for three weeks – so, when I get ill, I do tend to get a little worried and my parents and my husband tend to get a bit worried, but it [is all checked] very quickly, just in case it turns out to be something worse than actually it is because the doctor told me, with a chest infection, 'You can easily end up in hospital'.

They said to me, 'You may find it hard with some illnesses to get better' because, with pneumonia... trying to cough with a muscle weakness is very, very hard sometimes. People think I have a weak cough, but I just physically can't cough like I'm supposed to when I've got a cold or something.

Like I say, I'm not too worried until I have to get worried. I do worry that when I get older I might not be able to move around very much and be a burden on people – but you just have to see what happens. I'm not going to worry about it too much and, like I say, my main focus is my son – but I have a lot of help with my husband. He does things for Kane that I can't do, for example, pop on his clothes, carry him, and things like that.

I've got good family around me. All my family support me – my extended family – they know about it. They don't know everything about it, but they support me and understand some things I can't do. They are all very supportive; I've got a wonderful family. A lot of them do live close by, so I think I'm very lucky because a lot of people don't have the family support that I have.

I'm not worried about getting older with the condition worsening. Apart from that, my life is good.

On that note then, what are you looking forward to?

I'm looking forward to raising my son and being in a family with my husband, you know, and just having the best home life possible with our son – just hoping our son can do everything he wants to do; just encourage him, no matter what he wants to do, we'll help him with whatever he needs. If he wants to – I don't know – take up karate, or if he wants to do dancing even – whatever he wants to do, we're going to help him.

My husband and I, we're very good. We get on well together. We don't argue in front of the baby at all. We don't get into arguments any more. He is our priority. We're just looking forward to raising our son and having a good life together.

Great. Does it ever bother you that people don't know about myotonic dystrophy? I know that comes up.

With regards to people not being aware about myotonic dystrophy, I don't mind that they don't understand about it. They've never heard of it – but a lot of people, if they don't want to talk about it... There is another form of what you call 'muscular dystrophy' because a lot of people have heard about that and understand that it means muscles are weak. I don't really mind if people don't understand about it. If they want to ask me questions about it, they can. I don't mind that, but I'm not an expert. They can look on the Internet and probably find out more than I could ever tell them. I know certain bits, but I don't know the ins and outs of it.

Nobody actually seems to look at me, really, though. I've got a scooter now when if I'm going to the shopping mall like the Metro Centre. People don't look at me really, why is a young person like her in a scooter, because everyone has different things. You don't know what's wrong with them.

Like I say, yes, doctors, you know... Not a lot of doctors have heard of it – only specialists or anaesthetists, really, so it is a bit disappointing that they don't know about it but, unless a GP has got a patient with it, they're not going to know about it, really.

So, yes, a lot of people sometimes ask me, especially if I'm applying for a job, they sometimes say, 'Do you have a medical condition that will affect your job?', and I say, 'I do have a medical condition, but it won't necessarily affect the job I do'. And every time I go for a job, and I put it down, they say, 'Do you mind if I ask what it is?' And I go, 'Well, the Internet can probably tell you more than I ever could', but I say, 'Basically, it's a muscle weakness'. I think that might deter people from actually giving me a job. I know you're supposed to... 'Everyone is equal' and all this, but it does affect the way some people I know treat you.

I remember in my last job at the Department of Work and Pensions they used to say, 'You're not doing as much work as this other person on overtime'. So, if you don't do more work, they won't let you do overtime and now we know why with my muscle condition. There's nothing I could do about it at the time, but I didn't know, they didn't know – so what are you supposed to do if you don't know?

I wish more people are aware of it, really, because it may explain a lot. Other people might think, 'Oh, maybe I've got that'.

When I was in hospital when I was 17 with the operation I had, I had to have a physio come and see me to help me breathe afterwards because I had problems with it – and when I think back to that lady that came to see me – just the way she looked, I thought, 'I'm sure she must have had myotonic dystrophy, but probably without knowing it', because of just the way she looked. She had really heavy eyes, she was very thin, and just the way she was talking. I was just thinking, 'She looks like somebody who has it' just because, now, I know what people look like – the certain features they have.

Like I say, a lot of people don't know they've got it because it's hardly known about. A lot of people just thought, 'Oh, you're just like your dad in that respect' – you know, certain things that I did. But yes, that's why. It's actually something that nobody knows about. My gran – she passed away without knowing she had it. It's just one of those things, really.

I wouldn't want to have... You know, my kids are going to be fine – even being born. I wasn't too bad when I was born with it – but it is a hard decision to decide whether or not to have kids; to put them through the things they can get with it because they can have a hard life – especially at school when people tease them about certain things, 'Oh, you're stupid', or whatever.

I know there are some people who are incontinent with this which is awful at school when you wee yourself, because kids can be cruel – really cruel.

So, I do wish more people knew about it but it's just one of those things.

When you talked to me about the conferences – do you feel odd having something that a doctor can tell you have just by looking at you? You've got your family resemblance, and you've got your medical features. There's things that you said your dad and you have in common. Could you talk about that – what your thoughts are on people that you're not related to? If you see any resemblance or the way other people maybe take that?

I don't see anything weird about people that have this at all. I mean, everyone looks different. No one looks the same. You may have characteristics of your parents, obviously – but with people, I don't see them looking anything odd. Maybe they're very thin or just have a bit of heavy eyes, but I don't think they look weird. They just look normal to me. You wouldn't know there's anything wrong with them. You just think, it's just the way they're built, really – if you're thin or you're fat – it's just how you are, really. It's just the way you're born, the way you're built. I don't see anything weird about them.

Some people had a bit more difficulty walking and find it very hard to walk, but you just think, 'Oh, that person's got a bit of a problem with walking for whatever reason'. People don't have to know why. It's just, 'Oh, yeah. That can happen'. So, I don't see anything weird about the way people look with this at all. Sometimes you can tell. We all look similar, I guess, but no one else would know it because they think, 'Oh, they look alike'. Or they see someone and they think, 'They look like my friend'.

It's just, you think, 'Oh, that person looks like that person', but I don't see anything weird about the way we look, and there's nothing noticeable that any normal person who actually knows it's different... it's just who you are.

Great. Myotonic dystrophy affects so many aspects of your life; which aspect of it do you think affects or bothers you the most? If you could just change one of the symptoms, what would it be?

Well, it would be wonderful if I could find a pill that would just stop everything happening so everything would be good – but it isn't an issue. Because you're born with it, really, you don't think about that affecting you, because you've had it your whole life.

I have a problem swallowing tablets – big ones. I have to cut them in half, normally. I've always done that with round paracetamol, for example. I used to cut those in half. I couldn't swallow a whole one. And when I'm eating, I always have to have a drink – but that's just the way I've always done it. I don't think that's a problem because I've always had it like that. It's nothing that's happened new.

But the one that did happen new was when I got my stomach pains. I never had those as a little girl. It was when I started hitting puberty age – that's when those problems started happening and that's a thing that really does... I wish that could just stop. I wish I had a different stomach because it really can happen at the worst times. It can happen if you're out and about; if you're working with people; if you're on the phone... It can happen at the worst times and it can be really painful, and sometimes I'm brought to tears with it.

I think I have quite a high pain threshold, really, because I've had to deal with it all my life. I think I have quite a high pain threshold because this IBS – it really is painful and it's annoying and you also have the tablets with you like Diclofenac and Imodium, but also if you're out and about and you're having a drink – how are you going to swallow the tablets if you can't just swallow them? Some people just put tablets in their mouth and swallow them without water. So, to swallow tablets, I just dissolve them on my tongue, but it's not a good idea. Not all tablets can do that.

The only thing I would change, really, is probably that, and I wish I was a bit cleverer than I am, because that has affected my life – getting a good job or getting good grades – so I wish I was a bit cleverer.

I know my speech isn't perfect. I can't hear it myself, but when I hear myself back on a tape, I can tell. Some people find it hard sometimes to understand a word I'm saying. Also, when I try and tell people things, sometimes I go about it the long way round and they get sick of listening before you get to the point.

Sometimes I try to explain things to people or understand the [issues] of people, and sometimes I think, 'Oh, I don't know how to ask them; I don't know how to explain that'.

But like I say, the worst thing is probably my IBS because that really does affect day-to-day life, really, because it can happen any time. You just don't know when it's going to happen. It's really... It's one of the worst things and painful things I have to deal with, really.

Lucie, 34

Lincoln, 16 March 2015

Jackie: *Can you tell me your name and how old you are?*

Lucie: My name is Lucie and I am 34.

Can you tell me a bit about where and when you were born and brought up?

I was born in Carlisle in Cumbria, 1980, and brought up just outside Carlisle in a little village called Brunstock.

I went to a local primary school – a small country school. I think there were only about 40 to 50 pupils in total in the school. I went there when I was four, obviously stayed until I was 11, and then went to Austin Friars in Carlisle which was a Catholic school, but we weren't Catholic, but they allowed other multi-faith pupils as well. I was there from 11 to 18. I did my A levels [exams]; went off to university in London (King's College) and did a chemistry degree which took three years – so I was there '98 to 2001.

I then had a year out when I travelled. I went round Africa; did some volunteer work; worked to earn some money and then I went to the College of Law. I did a year to do the postgraduate diploma in law to qualify as a lawyer and I then did a year Bar vocational course to qualify as a barrister, and I then did a year as a pupil barrister – all in London. I joined Chambers as a pupil barrister and then as a tenant in 2004.

And you're now a barrister?

Yes. I'm now a barrister. As per most of the Bar, I am self-employed, but belong to a set of chambers that specializes in construction, energy, infrastructure, IT – so it's commercial disputes but with the subject matter being construction – big construction disputes, big energy disputes or infrastructure – so, railways, wind farms, oil rigs, IT disputes, IT systems that go wrong, some more general commercial work, but generally I focus on the construction and the energy sector and the IT sector, and that involves a lot of paperwork, a lot of advice, pleadings and ultimately advocacy in court.

Have you travelled a lot?

Yes, I travel a fair amount. We do international work; we do arbitration and a lot of the contracts and disputes that we cover are international disputes – so I've spent time, for example, in Dubai working on a project over there and we also travel a lot round the country to the various courts although a lot of the work, now, is focused on the technology and construction court in London. There are other court centres – so, last week I was doing a case in Birmingham. I do cases in Manchester,

Sheffield – so, there's a fair bit of travel and, obviously, I travel from here – home – Lincolnshire, into London – to Chambers – for work as well – so, it's a bit of travel.

OK. Can you tell me about your family life? If you're married and if you have any children?

Yes, I'm married. I married in 2011 and I have a daughter, Madeline, who will be two on Saturday and another one on the way who will be born, hopefully, end of September. My husband Russell is a doctor in Lincoln Hospital and we met at university when we were 19 – so, we've been together quite a while but only married since 2011. We lived in London as students and then for our working life up until November 2011, and then we moved here to Lincolnshire so that we could sort of – I don't know – get a nicer family home. We were both brought up in the countryside, so we've got the family home, the Labrador dog, and now the children – so, that's our life now.

OK. Can you say a bit about your wider family and home life?

Growing up, my family was just my mum and my dad and myself and my sister – just the two siblings. We grew up, as I've said, in a small village called Brunstock just outside Carlisle and that's in Cumbria, and that's where most of my mum's family lived and still do live – so we have my mum's parents quite nearby and my mum's sisters and their families – all my cousins. My dad's side of the family is from Middlesbrough, so we didn't see them as often, but they weren't a million miles away.

It was just a very normal life. My dad was a doctor; my mum a nurse – so we just lived fairly similarly to the way we do here – sort of house in a little village in the country, a dog (Labrador again) – so, I'm probably repeating the very similar lifestyle. I married a doctor, moved to the country, and got a Labrador – so.

And your position in the family? Are you the oldest or the youngest?

I'm the older sister – yeah.

Let's talk about myotonic dystrophy – what do you know about it?

Well, I've only really learned anything about myotonic dystrophy since my sister was diagnosed with it. I know that it's a genetic mutation; it's a genetic disease which runs through families; I know that it causes a wide variety of seemingly fairly disparate symptoms; that the symptoms generally stem from a muscle weakness or it's a disease that affects the muscles but that that can be anything from the heart (which is a muscle) to even your eyes, cataracts – I think it's one of those – well, I feel it's one of those diseases that's hard to describe because it can cause so many different problems for different people. That is basically what I understand about it, and what I've seen from my own family.

OK, so talking about that family, can you just talk about if you've had a test, when you've had that test, and how that came about?

Well, the subject was first raised just after my sister's first child Kit was born. He was born early. She actually went into labour when she was on a family visit here. We had my mum and all my female cousins and my sister – we all try to get together sort of once a year for a kind of girls' weekend, so we did it here and my sister went into labour while she was here, but she was only 27 weeks – so, she went off to Lincoln Hospital and had Kit by Caesarean then and there and was transferred to Nottingham. At 27 weeks we thought that was probably going to be OK but, actually, he didn't thrive and eventually, through that, they diagnosed her with myotonic dystrophy and Kit – so I was tested probably a few weeks after she got her diagnosis because, at the time...

Well, Russell and I just moved here; we were thinking about starting a family – luckily we hadn't already taken any steps in that direction – so I was fairly keen and Russell was fairly keen to know as soon as we could – so I was tested, I would say no more than a month or two after Sally was first diagnosed with it and that test came back negative – luckily – so, I don't have it and I don't carry it.

And how did you feel – what was that like?

Actually, I found out sitting on that sofa because, obviously, I'd been talking to my mum and my sister a lot because Kit, at that point, was in intensive care and at that point we had thought... [*very upset*]

I was speaking to my mum a lot because Kit was in intensive care and at that point we thought that, although he was doing not very well, it was just one of these things – he was premature and that, eventually, he would recover. I think, when I was told, Mum rang up and said, 'Basically, we've had some news; it's not very good' and then I think she just said, 'Sally's got myotonic dystrophy; Kit's got myotonic dystrophy'; and I think she first sort of explained what that meant for him and said... I think she probably did say at that point that what it meant was that he probably wouldn't survive. That was probably the initial thing that I was thinking about, and then she said there's another problem. The bad thing is that it is genetic, so there's a possibility that you could have it as well – so I'm not sure what I really thought at that point because it was mostly this sort of going from thinking, 'Oh, well – Kit will be fine; it's just a matter of time', to thinking, basically, 'Kit is going to die', and then, on top of that, being told that I could have it as well and I think, probably the first thing I thought (because my husband was sat next to me and we had actually been talking about starting a family), the first thing I thought, I think, was 'Oh dear, does that mean that we're not going to be able to have children?' Probably more so than thinking, 'Oh, God, I've got myotonic dystrophy. Are we going to have babies that die?'

Russell sat next to me so I told him and he knew more about it immediately because he's a doctor so he knew a bit about it. He immediately started asking me to hold his hands; squeeze his hands and let go; and started going through various things saying, 'Well, you've never had a problem with this; you've always done really well academically...', I think, possibly to reassure himself as much as me, starting ticking off a tick-list of things to try and reassure us both that I probably didn't have it.

I think by the end of that, and I think after a day or two of mulling it over between ourselves, I was fairly convinced that I probably didn't have it just because... I mean, Russell then looked it up; he looked at some medical journals online; he went through all of the different symptoms – the way it displays itself and we, one by one, went through all these things and said, 'You don't have any of those; you don't display...' In fact, my characteristics were almost the opposite. I've always had a lot of energy; I've always been really full of drive; very academic; I don't like sitting still; I'm not one for sleeping much; I've never had any balance problems; I've never had any health problems at all; my eyesight is good – so, I think, over the course of a couple of days I quite quickly reassured myself and he reassured himself that it was very, very, unlikely that I had it although, obviously, we still wanted to be tested – so, I think the focus was very quickly, really, onto my sister and Kit – or really on Kit because that was the immediate problem.

He was in intensive care and we just got immediately thrown into having to go and see him, and then he died shortly after so I think that was, to be honest, that was the main thing I remember – mostly having to deal with him having died, and my sister, and then I think it was only after Kit had died and after all that had settled, I had my test, and that came back negative but at that point we were fairly sure it would and that, after that, it became more about my sister and more about worrying what effect it will have on her.

She'd always been... It explained a lot. We all sat down and, 'That's why Sally is how she is'. You know, because she'd always been basically the opposite. Nobody could believe we're sisters. They were like, 'Everything about you two is the exact opposite' – so I was academic, she wasn't; I was full of energy, she wasn't; I wanted to go off, leave home, travel, be independent – she used to say she didn't ever want to move out of Mum and Dad's house. So it explained a lot about her and she's had some health problems, but nothing hugely severe, but I think once she had Kit that took it out of her – so then it became very quickly about Sally and her health problems and trying to... I don't know... I think that's what it's been since then – it's trying to figure out all the things that are wrong with her; what that's going to mean in the future; worrying about that... Yes, I think in terms of my own personal...

For me, apart from a few days of worrying about, 'Do I have it or not?', which was a very short period of time, it was mainly worrying about Kit and then worrying about Sally as opposed to concern for myself.

Can you describe your sister to me?

Well. Sally's appearance has changed quite a lot. When she was younger, she was a really pretty, chubby little girl. We used to call her 'Michelin Baby' because she had the rolls of fat. Her face has changed quite a lot because she had a jaw operation when she was 18, I think, because her jaw was twisting in one direction. She couldn't bite her teeth together and that got worse and worse over time. When she had that operation, her face changed quite dramatically – so, as an adult, she doesn't really look as she did when we were growing up.

Looking at her now, and looking at people with myotonic dystrophy, I can now see that she has features of myotonic dystrophy, but I had always previously put down the way that she looks now to the fact that she had this jaw operation which did leave her features looking different to how we had grown up, and a lot different to me.

She used to be a chubby little kid... She is very thin now. She is a lot thinner now than she used to be. She was always quite a laid-back, lazy child, basically! I remember Mum always having to get her out of bed; run her bath for her. This is not just as a small child – this is a teenager; this is into adulthood when she was still living at home in her twenties. My mum would literally be getting her out of bed; lay her clothes out; putting her bath on; putting her in the bath... She never had any sort of get-up-and-go which was polar-opposite to me because I never used to sit still and I was always doing things and always running around. She was always very shy. She had a speech impediment when she was younger which we thought was due to her jaw, so she had speech therapy when she was little but that meant that she would never go into a shop and ask for things, and I always had to do everything with her – so, we went to classes (sort of, gym classes – gymnastics and things outside school and riding), but I was always going with her and I would always do the speaking. She was very, very shy. She was always very dependent on my mum and my dad – but particularly my mum.

We'd always just said, you know, 'Sally is a bit lazy'. She always wanted Mum to do everything for her. And, again, I was the opposite. I preferred my parents just not to do anything for me. I could do everything myself and go off...

She didn't like school; she didn't like exams. She was diagnosed with, I think, dyspraxia as opposed to dyslexia which, at the time, we thought that explained her performance at school because she was never academic; she never did well in exams – but luckily (and I always thought this was just luck), because I was always doing really well in exams, and everything I did I always did well at and excelled, and she would do exams in the same school as me and she would always do really badly, and I used think it's really lucky that she doesn't have ambition – it didn't really seem to bother her. I used to think, 'Oh, doesn't she get annoyed that I always do well in things and she never does?', and she didn't really seem to get that annoyed about it. I would have really cared if I came last in an exam – I would have really cared. She never really seemed to be that bothered and I always thought, 'Oh, that's lucky. She doesn't want to go off and be a doctor or a lawyer'. What she wanted to do was stay at home and be with Mum and Dad and just have a nice job near to home and family. And I thought that's good, because otherwise she would be really annoyed at me, or really resentful.

She's always been very kind, but to the point of naivety, I think. She is not cynical in the way that I would be, or in the way that I am. She is never very streetwise. I think she is a very quiet, very kind, and very gentle person and I think, probably, we all take, or have taken, advantage of that in the past! And she is very easily led and, again, I'm the opposite of that. I tend to err on the side of cynicism and definitely not easily led.

What else to say about her? She seems to be very content in her own world. She had some odd talents – for example, she loved watching TV and videos as a kid. She would watch the same videos over and over and over again, but she would notice continuity problems in videos. Maybe it is something about the way she perceives things, is more visual or something, because she always had problems with numbers and she always had problems with letters, but she can watch a video and she would spot continuity issues and go, ‘She didn’t have her handbag on the left shoulder in the last clip’, which we would never see – so, she’s got these sort of odd talents and things.

Yeah, and I think she’s just a quiet, very homely, person.

Do you think she’s happy?

Um, it’s hard to say. I think she is, but, because we’re so different, I find it hard – I’ve always found it hard to judge her life or how she must feel about it because I probably impose my own sensibilities onto it and I think, in her position, I wouldn’t be happy because I wouldn’t be happy...

She didn’t get to go to university; she used to work for the Department of Work and Pensions in a job that she really enjoyed but I would have considered as admin; and she didn’t get to travel; she’s never really been places, seen things; she doesn’t... You know, as far as I can see, she doesn’t have a full, exciting life. She very much stays at home; she watches TV; she is with her husband.

She’s got Kane now, which is good. I think that makes her happy, but I’ve always found it difficult to see how her life is fulfilling in the way that she lives it – but then, I think I’ve just had to accept that we’re really different people and that, actually, that life, and being near Mum and Dad and seeing them every day, and seeing the family, and being at home quietly and having that sort of job – maybe that does make her happy. I hope so. I hope so – but I don’t know. She doesn’t really express either great happiness or great sorrow – generally. I mean, even... Yeah, I don’t think she expresses her emotions very obviously.

I think she has been very unhappy. She’s obviously very unhappy about Kit. That would have made me extremely unhappy – but I don’t know if it’s part of the condition that she has, but she seems just to accept things in a way that I wouldn’t. She just has accepted all these things that life has thrown at her, whereas I would be very frustrated by them, or very angry or upset. She seems to have a very kind of, ‘Oh well – that’s life’, and just gets on with it – that attitude. So, I don’t know.

Hopefully, I think that now with Kane, I think she is happy now with Kane. Well, I hope she is, anyway.

Can you talk about growing up together? Was there any rivalry around sharing clothes or shoes or anything?

I don’t remember sharing clothes so much... I suppose we did share clothes, but only in the way that they were passed down from all our female cousins because we had four female cousins on my mum's side above us, so we'd get a black binbag of clothes from them which were normally about five to 10 years out of date in terms of fashion – but then, who knew here in Cumbria, anyway? Then I'd wear them and then my sister would get them. I don't think... Very rarely would she try and borrow my clothes. She was bigger than me. Growing up, she was taller. She was generally bigger. I was really, really skinny; she was always more curvy – so, there wasn't actually much... There weren't many points in time where we could have actually shared clothes anyway. Now she's probably closer to my size, but there wasn't really much rivalry either.

We fought. We did fight a lot. She was actually stronger than me. She was more physical than me, so I would sort of verbally... It would be verbal abuse from my end, and then she would come and beat me up and I would have to lock myself in the bedroom. I'd lock the door and she was like banging on the door trying to get to me. She knocked my two front teeth out once when we were little.

I don't think there was rivalry in the sense of teenage rivalry because I had a very different set of friends and a very different set of interests and a different lifestyle to her – so I don't actually remember any teenage rivalry. I don't remember much clothes-sharing or shoe-sharing – no.

What are your thoughts on ageing? Do you think you have aged, seeing yourself 10 years ago?

I think I must have aged. I don't feel in myself a lot older, but then I look at what I'm doing now – so, for example, I now spend weekends shopping for antiques and I can remember, when I first moved to London, for example, moving into the first house there and my parents saying, 'Oh, do you want some of our furniture?', and we said, 'Oh, God, no! Not that stuff. It doesn't match!' No – I want all my stuff from IKEA, all modern and now I'm reading gardening books and going to gardening shows and having the garden done – so, I think the way that I can perceive myself ageing and judge myself ageing is when I look at the things I enjoy doing and the activities I do now – things like going to antique shops; going to garden centres; saying things at the weekend like, 'Oh, wouldn't it be nice to go and see that National Trust place down so and so because that has a really nice house and gardens?' And then just thinking, 'Oh, God – that's the sort of thing I used to think is so boring; my parents are so dull'. I can see myself turning into my parents!

Ten years ago – what was I? – 24; I was in London, so I'd be wanting to go out to pubs and meet friends and go clubbing and now the *last* thing I'd want to do is go clubbing. The thought of that just makes me go, 'Uh!'

So, yes, I don't feel physically like I've aged. I've not yet sort of hit a point where my body is not letting me do the things that I want to do – so I can ride and I can run and I can do all of that, so physically I don't feel I've aged hugely.

I suppose mentally I must be more responsible now – I assume. Sometimes I feel like I just act like I'm more responsible now – but yeah – I think it's lifestyle. My lifestyle tells me that I've aged as opposed to anything else.

If you had to tick a box with your age, what box would you associate yourself with? Like, young, middle-aged, older?

I still feel like I'm a young person. I think, before I had Madeline, I felt like I should still be in my twenties, and the thirties had just kind of happened, but that had only *just* happened. I'd probably be happy to put myself in the age bracket that I'm in now. I feel probably more early thirties than 34. I'll be 35 this year. That feels like I'm edging a little bit towards 40. It's hard to tick 'young person' when you've got a child, I think, because now they're the young person and you're a mum – so, I would find it more difficult to describe myself as 'young', but I certainly wouldn't describe myself as 'middle-aged'. I don't know if there is a word for the bracket that I fall into, to be honest.

Thinking about your wider family, who do you think was affected the most by the diagnosis?

I think, in terms of wider family, the most immediate effect on the wider family was, once my sister was diagnosed, they then wanted to check if it was my mum or dad and they didn't even bother testing my mum because it was fairly obviously my dad – obviously, with hindsight – although no one had ever suspected there was anything wrong with him other than just normal old-age related...

But once it was my dad, then the immediate effect on the wider family was that he had two sisters, one of whom died, but both of them had had children and, in fact, one of the sister's children had had children – so they had to be told and all tested. But, again, I think we'd all figured out that one of my dad's sisters probably did have it because they look like twins. They are pretty much identical and her youngest child had had medical problems from an early age and she had been diagnosed with some sort of syndrome from an early age which I think now, with hindsight, she actually has got myotonic dystrophy. She's been tested.

So, two of my cousins had it and one of my aunts – but that's where it stops for our family. My dad's other sister had died but her daughter doesn't have it. I'm not even sure she was tested because she'd already had lots of children and her children had children and there was no sign of it – so I'm not even sure if she was tested.

So, with the wider family, I suppose there were those implications. Neither of my auntie's two daughters that have it have had children – and her son doesn't have it.

In terms of the effect on the people who don't have it in the family, I think it's obviously most difficult for my mum because she was really close to my sister. My sister's always relied on my mum for everything. My sister rings my mum, without exaggeration, at least five or six times a day, possibly more – like, when my mum is staying here, the phone rings at least four or five times a day and we all say, 'Oh, that will be Sally'. She speaks to my mum *all* the time; my mum does everything for her. So, I think it's hard for my mum and, obviously, with Kit it was really hard for my mum.

I suppose my mum was having to look after my dad a bit, but I think she'd already accepted that, because he already had his problems and that was just old age and she just thought he was just a cranky old man, and she just sort of put up with that.

I think a lot of the burden of looking after my sister falls on my mum and now with Kane my mum has to take on the burden of helping with him a lot. I think she's the one that worries the most about what will happen to my sister... I don't think she worries so much about Dad, and I don't think Dad worries about Dad. My dad is 70 next year. He was in his late sixties when he was diagnosed. He already knew he had these health problems. He didn't know it was because of myotonic dystrophy, but he knew he had them and, to be honest, I don't think there was anything he didn't expect as a consequence of old age – so, I don't think it's had much effect between Mum and Dad in terms of worrying about him. He is as he's always been. I think it's been my mum and my dad worrying about Sally; about how it's progressed quite quickly with her.

Before she had Kit, she was reasonably healthy. She wasn't a picture of health, but she didn't have anything seriously wrong with her and then, as soon as she had Kit, I think that took a lot out of her and she went downhill very quickly and then Kane – although he doesn't have it – has taken a lot out of her and she's gone downhill again very quickly, so she's actually, since 2012, she's gone from being just a normal, reasonably healthy daughter who, OK, required a lot of support from my parents because she was shy, because she wasn't that academic, because she always had problems managing her life – but now it's that, plus they've constantly got to worry about her health problems and the uncertainty as to, is it going to stabilize; is she going to get a lot worse; is she going to end up in a wheelchair; what are they going to do about that? So I think they probably worry... Both Mum and Dad probably spend a lot of their time worrying about that and every day dealing with the consequences.

What emotions do you associate with that? People say they feel guilty. What kind of feelings do you associate with it?

There's a lot of different emotions. The strangest emotion which, I think, my sister had as well, was a form of relief in that... Well, not relief... I don't know how to explain it. Because there was suddenly an explanation, it was like a lights-on moment because it had always been puzzling as to why Sally was the way she was, and frustrating. I'd always found it quite frustrating that she would depend on Mum and Dad for so much and that she would never get up and do things and I'd say to her, 'Just do it yourself, Sally. Mum and Dad spend their whole lives running around after you. Just do it yourself!' There was that frustration.

I think she found it frustrating. Mum said she used to say to Mum, 'Why am I like this? Why am I lazy?' And you see, my answer always used to be to that, 'Well, don't be lazy then, Sally. Just get up and do stuff'. I think to suddenly go, 'Oh, that's why!', I think Sally found it a bit of a relief. I think she even said that – to finally have an explanation for the way that she was and there was...

But the flipside to that was, OK, so now we know why she is the way she is, but then it's something which actually has got medical consequences as opposed to previously, with her, it was just more behaviour – more the way she was – and so there was also... I don't know – there's just worry. I suppose that's probably the other main emotion – it's worry, because you don't know what the outcome is going to be for her. Nobody can say – it may be that she will live to the age of 70 and have some disabilities, or it may be that she won't. The thing that I worry about is that she'll die before my mum.

So, it's worrying, but the problem is, you just don't know. It's quite frustrating as well because nobody can tell you like they maybe could with other illnesses. They can say, 'Well, this is the way it will progress and this is the diagnosis and the prognosis'. So, worry...

I felt bad that, I suppose, that I'd always got at her for things like being lazy. I'd always said, 'Oh, you shouldn't make Mum and Dad run after you so much and blah, blah, blah. Just get off your butt and do stuff, Sal'. So I felt guilty that I'd always sort of pushed her like that, or treated her like that and just thought she's just being lazy when, actually, it's something that she couldn't help; it's just in her and I still struggle with that.

She's meant to do things like physio to keep herself healthy. She had pneumonia recently and she was in hospital and the nurses were saying to her, 'You've got to get up and you've got to do a walk once a day' – and she wouldn't do it; she wouldn't do it – and I still get frustrated and I still want to say, 'For God's sake, Sally. You've got a child now. You've got responsibility to just get off your butt and do some physio and try and be more healthy. You've also got responsibility to us because we're the ones who's going to have to sit here watching you deteriorate and potentially die at the end of the day. Do something about it! Save yourself! There are things you can do'.

So, I still do find it difficult not to treat her like that. My mum has always been better with her and always been – I don't know – understood her personality more. So, I find it frustrating because I still want to say that, but on the other hand I'm thinking, 'Well, maybe it is...' I don't understand it, basically. I don't understand what it feels like and what the syndrome is mentally.

I mean, the lethargy thing is the bit that I don't understand. The medical conditions you can understand scientifically – that she's got a heart block – that's a physical thing, but the mental things like the fact that she won't do physio. If it was me, and I was looking at having a condition where, potentially, I'm going to deteriorate and be in a wheelchair or get pneumonia or even die from it, I would be doing everything I could to stop that happening and she doesn't. I don't know – she doesn't understand or does understand and doesn't... The mental bit I find very difficult to comprehend – this lethargy and this apathy.

So, yeah – I think worry, frustration, the lightbulb moment that explained so much, guilt for almost treating her that she was just being lazy. I don't know if I feel guilty that I haven't got it and she did. That's difficult because I don't think...

I'm quite a logical person and I think there's nothing I can do about that; in fact, it really wouldn't help if I had it. It would have made things 10 times worse for my parents if I'd have had it – so I'm glad that I didn't – for myself, obviously, but for my parents. I don't think they could have coped with two of us. They put all their energy and all their time into Sally. I'm not sure... Well, I suppose they would have coped, but it would have been really difficult.

So I think, yeah – those are the emotions. There's a lot of them. But also, you just resign yourself and get on with it, don't you, really? There's nothing you can do at the end of the day. You've just got to get on with it and hope for the best.

I suppose it probably is quite... It maybe is a common thing because most siblings wouldn't be as different to each other as that. You know, a lot of siblings are not as polar opposite as that and I don't...

One of the things that I have wondered about – probably just more recently once we got over the shock and now it's all settled down – is I have started to wonder about what she really would be like without it. I know that is probably a pointless thing to wonder about because we'll never know – but if she didn't have myotonic dystrophy, what would she have been like? Because, obviously not everything about her is myotonic dystrophy, but it's difficult to know how much is, particularly because, as I say, the physical things – put them to one side: the heart condition and all of that; the fact that she limps now and the muscle weakness – if you put all of that to one side, the personality traits – so, if she didn't have myotonic dystrophy, would she have been academic because my dad is very academic, my mum is academic, the family generally – my gramps is a chemist... So I wonder, because, genetically speaking, without myotonic dystrophy, she should have been reasonably clever with the parents that she's got. So I'm thinking, would she have been or would she not have been without it and, you know, both my parents have always been energetic and again, I think, without it, would she have been more energetic; would she have been a go-getter; would she have... You know. Or would she still have been a homely, shy person? I don't even know how to begin to answer that question, but it's almost like you think, is there another sister that I would have had if she hadn't had myotonic dystrophy, and what would she have been like, and would she have had a much better life?

So, I think, obviously, elements of her life would have been better; her health would have been better – but, I don't know... I just find that an interesting question. Would we have been more similar? Because I think, because of the condition, however different we would have been has been magnified tenfold, and so we're literally at opposite ends, and most siblings – I mean, look at my husband and his sister and although they are very different, they're not as different as we always were and, also, as I say, we never really, because we're so different personality-wise, I've never really understood her and she's probably never really understood me – so we got along on a level. We never were like really close because we just weren't into the same things so I wonder again, if she hadn't had myotonic dystrophy, would we have been closer; would we have been more similar?

I think that's an interesting question and I've only got one sibling, so I don't have another sibling to compare it to either. I don't have a sibling that doesn't have it so that I can say, 'Oh, actually, yes we would have been because there are two of us here that are...' So, I don't know, but that's something that I probably only more recently started to wonder about.

In relation to that, is there a sense of loss there?

That's difficult because if you feel you've lost the sister that you would have had, if you had had her, you've lost the sister that you do have – so, I think I feel it more for her. I feel how much better her life would have been if she hadn't had it as opposed to me thinking, 'Oh, it would have been much better for me to have had a sister without because maybe we would have been closer; maybe I would have understood her more'. I don't feel that I lost that sister because I've got the sister I've got, and regardless of how different we are, I love her as much as you love any sister. Now that I have a child I think it would be a bit like thinking, 'What if I had a boy; what if I had a different child?' You don't even think that. You can't even conceive that because the child you got is the one that you love and you wouldn't be without them – so, I don't feel it in that way. It is more of a wondering.

I suppose, as well, because I don't know the answer to how different she'd be... So, if someone said, 'You could do this now and she wouldn't have myotonic dystrophy and she'd never have had it', would that then just be a stranger turning up and it wouldn't be her anymore because I don't know how much of her personality is defined by the fact that she's got myotonic dystrophy so you wouldn't want to just click your fingers and your sister would be a different person because then you would have lost your sister that you've always had.

I think that is a really difficult question to answer and I think unless I knew the answer, unless I knew I could click my finger and she'd still be Sally, but she just would be more healthy and have more energy and be able to have a better life, then maybe I would do it if she'd still be her, but I don't know if I did that if she'd just be a completely different person and then I would have lost *my* sister – so...

I don't think it's possible to answer the question if I feel I've lost something because I don't know what it would be that I've lost.

If you had a magic wand and could change one symptom to take away from a person with myotonic dystrophy, what would you choose?

I think if I could exercise that power in six months' time, then the logical part of me says that what I'd do is, I'd have her fully tested; I'd find out what the bit of her condition is that is going to be the most severe for her (so, whether her heart condition eventually that's going to...) – so, I think my logical brain is telling me you would take away the symptom that is most likely to kill her in the end – so, if her heart condition is going to deteriorate, or if it's her legs and she's going to end up in a wheelchair – that's the logical part of me.

My gut reaction, if I had to just do it now and I didn't have any... Because at the moment I don't know the answer to what the most serious thing about her condition is for her. I know her lungs are not very good. But if I had to just do it on the spur of the moment, I'd probably take away the apathy and the lethargy, to be honest, because I think that whatever's going on with her, what she really needs to do is to be able to motivate herself; to help herself; to exercise; to stay healthy and I also think just for her life in general.

I just said to my mum the other day because I was feeling ill with morning sickness and I said, 'God, that's my worst nightmare to just have to sit around all day watching telly and not do anything. God! What sort of life is that?' We were talking about Sally and the fact that she needs to get out and do some physio and she needs to get out of the house and start walking and I said, 'God, what sort of life is it, Mum, that she sits there at home, and she watches telly, and she doesn't *do* anything?' You know, however long or short, or whatever happens in her life, at least she should *do* something with it and have some good memories – so, I think if I had to just do it now, I would take away the lethargy and the apathy so that she had more get-up and go.

But then, on the other hand, I suppose, maybe I'm over-thinking it, but then, I suppose, what if that took away the acceptance as well and that, suddenly, she had like my personality wanting to do things, but she had this thing that limited her, or she would suddenly realize how serious it was and that she physically couldn't do things and maybe this apathy and this lethargy is helping with that because it shields her from the frustration that I would feel in that circumstance. She doesn't feel that frustration.

It is actually a very difficult question. Yeah, I think, without knowing which part of the condition is the most serious... It's a really difficult question, actually. Yeah, maybe I wouldn't wave the wand to get rid of the lethargy. Maybe I wouldn't because, actually, if you think through the consequences of that...

It could well be the muscle weakness, actually, because she struggles to pick Kane up a lot and I think, actually, the one thing she would really love... I think if she had the choice, would she say that she would want to be able to pick him up, and be physically more able with him? Maybe I should just ask her, and wave my magic wand to get rid of whatever symptom she wanted to get rid of!

To be strong.

That could be it. It's difficult, though, because I impose my own sensibilities. I think, if it was me, the worst thing for me would be sitting round all day doing nothing – so I'd want to get rid of the lethargy and the apathy, but you think about it and you think, maybe that's a double-edged sword and maybe that's not what she would want.

But certainly, if somebody told me her heart condition is going to deteriorate, you'd obviously get rid of the thing that's going to most severely impact her health but nobody seems to be able to tell you that. Nobody seems to be able to say to us,

'This is the thing that's going to do the most harm health-wise or in the long term', or, 'This is the thing that will eventually not be survivable'. I don't think anyone, not even Professor Lochmüller has been able to say, 'This is the one element that is the most serious for her'.

I think that's what's the thing that eventually frustrates us. It's not like saying, 'You've got cancer'.

Yeah, exactly.

It can be so many things. It's not always the thing that patients want cured which is the same with medicine. For years doctors were fascinated by the sticking hand (when gripping) thing, but it doesn't really bother the patients that much. The tiredness has a far bigger impact on folks' lives.

It's not just apathy and lethargy. It's the acceptance though that goes with it – the *laissez-faire* attitude. If it's just laziness, that's just like, 'I can't be bothered to get out of bed, but I'm still sitting here being very distressed by the fact that I've got this condition and I'm frustrated by various things', whereas – I don't know – maybe she is, but doesn't express it. It's not *just* 'can't be bothered to get out of bed', or an inability to see consequences or plan ahead... That's why when I said I'd lift the apathy and lethargy, it suddenly hit me, if you did that, and she suddenly understood the consequence, then would that be worse for her because she would be sitting there going, 'Oh, I've got this thing which means I might die; I might be in a wheelchair...', and she'd then think ahead and get, as I would get – I would be thinking ahead and getting really depressed about all the things that are going to happen – maybe it's shielding her from that.

Again, it's not just academic; it's not just that she can't do letters or numbers. I think she finds it difficult to understand logical concepts and consequences and there's a social – not like autism, but it's like a slight social unawareness, an inability to interact completely normally with friends and with people. How to describe it? A social awkwardness or a social difference that comes with it – so, it would be interesting if they could get to the root of that and how it all combines to cause this because there's not actually that many illnesses that affect your behaviour as well as your physical being as much as this one seems to do. Yes, it is very interesting.

There are patient-support groups. Have you ever been to any of the patient-support meetings?

No, my sister and my parents go to the one in Nottingham. They go every year.

Do you know other people who have it?

No, not outside my own family – no. My mum describes going to the conferences and coming back and saying, 'Oh, God, there's some really disabled people there', and people where their children have died and things and part of me doesn't want to go and see all of that because you would then just be thinking, 'Oh, is that what's going to happen?' I think I understand probably as much as I want to or need to from Russell, and from reading medical journals and I've read a lot of information on the Internet about it. Maybe I should, but at the moment I'm leaving that to my parents in getting involved. They come back and tell me what Professor Lochmüller said and they come back and tell me what was said at the conferences but I... Yeah, I just don't know if I want to go there and see.

Also, because part of me thinks, is that really relevant, because everyone is so different. Are you not just going to be scaring yourself unnecessarily because, as far as I can tell, every single person who gets it, although there are commonalities, the physical consequences can be so diverse that you just have to take the person that you've got – the family members that you've got and take their particular progression through it and their particular symptoms as they come along as opposed to worrying too much about seeing other people that are very badly affected?

Yeah, I think, consciously, I've probably decided not to go to the conferences.

Yes, I find that very hard, and my mum did – but then, you also meet people who are much better, who are holding down good jobs.

The information is good, but sometimes... I kind of pushed it to find out more about the condition but I can see why you want to keep things in the present, think about maybe five years away but not more.

Also because I think it's so diverse. It seems to me that one ends up treating all the different symptoms of it, but treating them as symptoms – so, if you have a heart blockage, that is a heart block; if you've got lung problems, it's lung problems – but there's no kind of one pill that you give someone with myotonic dystrophy that just treats them. You can't *treat* myotonic dystrophy. You treat the things that it causes – so my dad is taking pills to stop him falling asleep; you treat each individual symptom.

It's a genetic thing, so I guess it's not something they're going to be able to just go, 'We're going to cure it now'. It's not one of those diseases. Maybe, eventually, when they've got genetic cures, but that's some way down the road. When they can start curing genetic things, then that's it. Lots of genetic diseases will be cured, but I just... I don't know...

Maybe, eventually, I will want to understand more – generally – but at the moment I just think... When my mum and dad went, my mum was quite shocked as well, like your mum. I think that brought it home to her how serious it could be. I think she just wanted to pretend it wasn't all that bad – especially as Dad had it and he's nearly 70. He's not healthy, but he's not particularly decrepit for a 70-year-old. You wouldn't look at him and think he's any worse than your average 70-year-old.

I think some knowledge is good, but I don't think you need to unnecessarily scare yourself. There's no point either because there's nothing you can do; there is no pill you can come up with; no cure to be given – so... But it's good that Mum and Dad do it because they need to encourage her to do the things that she can do –

like physio, going swimming. I keep going on and on at them, 'You've got to book her some appointments'. If something is booked for her, Sally will do it.

She did have a job. She actually went to do her job. She would get up early; she'd get up in the morning; she'd get herself to work. That's the thing I keep saying to them, she needs to not use it as an excuse as well, because now she often goes, 'Oh, I've got myotonic dystrophy', and you go, 'Oh, well, but Sally...' I think if they book her things that she's got to do – so, if they booked her a physio appointment, she would turn up. She will do things when she has to do it, so I keep saying, 'That's the way you've got to manage it; you've got to force the issue by booking things', and then she'll actually go and get them done.

But – you know – I have to leave it to my parents because she doesn't listen to me, anyway. The only person she'll listen to is my mum. She's quite stubborn.

I think stubbornness is very common.

Really?

It is in my family.

That's the frustrating thing. If it was me, these are the things that I would do – but then you think, if it was me, if I had it, I probably wouldn't do those things because I'd probably be sitting there going, 'Can't be bothered to think about that'.

That's the really frustrating thing, isn't it – that you can't put yourself in their position; or to try and not impose your own – the way you would think about things – onto them because you would just... But I find it really difficult.

It's like an empathy thing. It's really difficult. You can be sympathetic; you can be upset – but empathy which means understanding the way someone feels – I actually don't think... It's like trying to imagine the universe and infinity – you can't do it. I don't think I can imagine myself into her brain – so I think it's really difficult to communicate. I just say, 'For God's sake, Sally, get up and do this', and I don't understand why she won't.

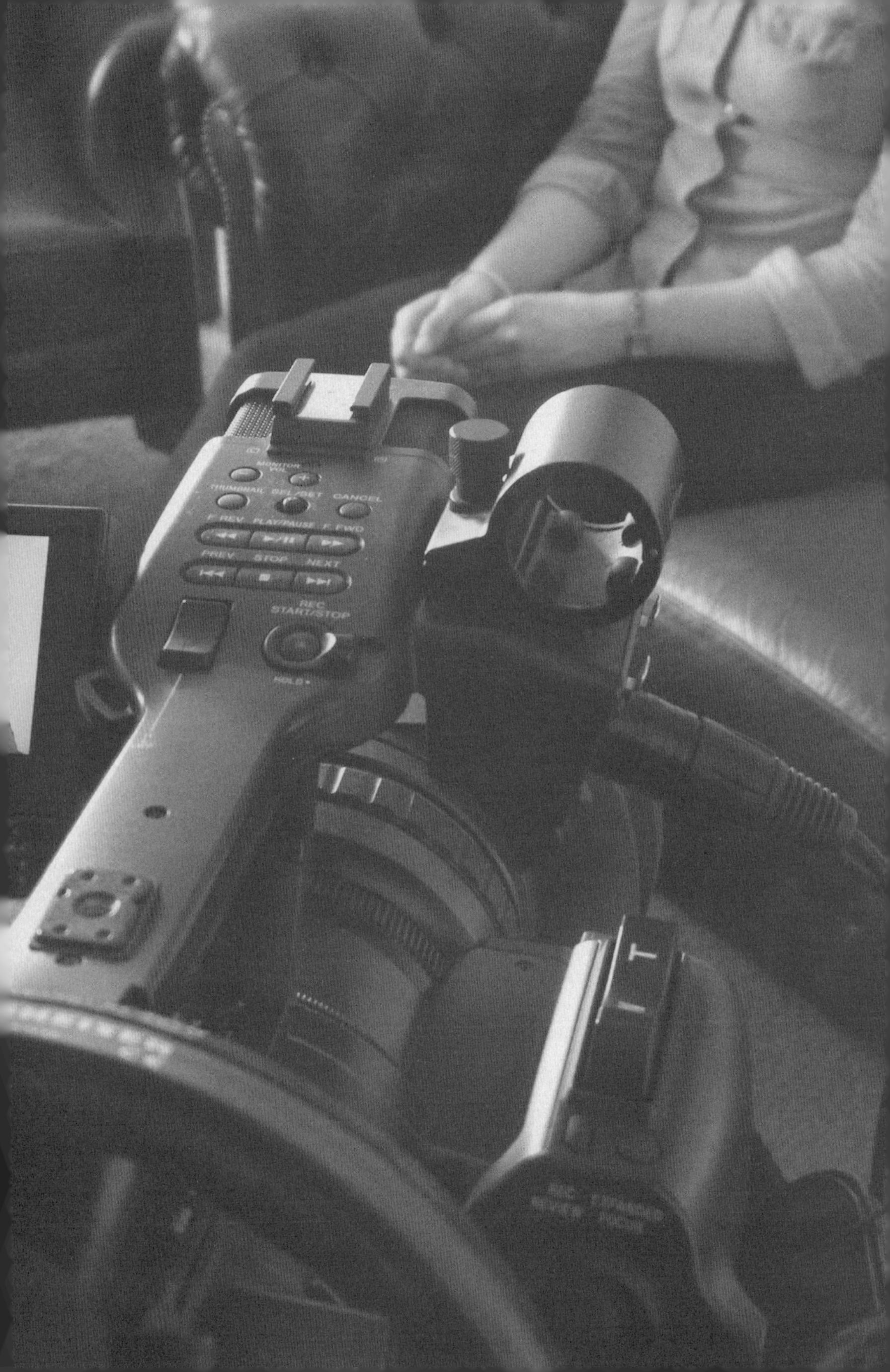
MONITOR
THUMBNAIL
CANCEL
PLAY/PAUSE
F.FWD
PREV
STOP
NEXT
REC
START/STOP
HOLD

Claire and Laura

Claire, 31

Age at diagnosis: 22
Northern Ireland, 6 February 2014

Jackie: *We start with some really simple questions like – can you just tell me your name?*

Claire: My name is Claire.

If you could say how old you are and where you were born?

I'm 31, and I was born in Belfast.

Could you tell me about your siblings?

I just have one sibling and that's my sister Laura.

And how old is Laura?

Laura is 29 – I believe – 28, 29.

If you could talk about your childhood – so, where you grew up, and where you went to school?

I grew up in Belfast in our family home. I went to school – just a local primary school, and then we both went to an integrated secondary school in Belfast – one of the first integrated schools – so, we had a pretty tempestuous childhood – I suppose we can say that!

And did you go on to college or university after school?

I went to college for two years and did beauty therapy and then I left and went to work abroad. Laura stayed at school and went on to college to do A levels [exams] and then she went to university.

What do you do now?

I work in a nursing home as a care assistant, and I basically just help people with their day-to-day lives in a nursing home – so, it's good fun; it's hard work, but I really enjoy it. We have 32 residents in our nursing home and my job is to get them up, get them ready, get them washed, dressed, get them something to eat, and just their basic day-to-day care throughout the day.

I'm going to ask you a few questions about myotonic dystrophy – some basic things. Could you tell me what you know about myotonic dystrophy?

I know myotonic dystrophy is a degenerative muscle disease. It causes wastage of the muscles in the body. People are affected very differently, from one extreme of not being able to walk, or being born very disabled, to degenerating throughout life – so, small things happening and you could end up completely disabled or very mildly affected – just depends on what stroke of luck you get, basically.

Could you tell me when you and your sister were tested?

Yes, we were tested when I was 22 – I think. I can't really remember what age I was. We both went along together to go and see our geneticist, in the city, and we both had our blood tests done and our ECGs and different things, and then we were called back about three weeks later, I think, because we both again went down together and that's when we got the diagnosis that I was positive for it, and Laura was negative for it.

It was quite an emotional day because I was fine, and Laura bawled her eyes out. She was the one that was upset. It was not her, and it was me – so it was a bit of a strange day.

And how did you feel about the results?

I was fine – I suppose because Mum and Dad talked about it. I have an auntie who has myotonic dystrophy and I just sort of felt, 'Well, that's just the way it goes'. We knew one of us could have, both of us couldn't have – we didn't really know. I was actually... I just embraced it, and I was really fine about it, and Laura was really upset.

Could you say if you have noticed that myotonic dystrophy has an effect on your wider family?

On my wider family – yes. I had an uncle who had myotonic dystrophy. He unfortunately died when he was very young. His was quite prominent with his speech and with his walk. He walked very stiff legged, and his speech was very hard to make out and my aunt as well. She was the same. When I was young she babysat me, and my uncle as well. It was slower, I think, with Elizabeth – more degenerative – until she had a little boy and he had very profound myotonic dystrophy at birth. He sadly passed very quickly, and that's what took its toll on Elizabeth, I think, more than anything else. She started going downhill then after that.

Have you noticed any symptoms in yourself?

Well, I don't know. There are symptoms that are classic myotonic dystrophy, but also could be completely irrelevant.

I have a cataract in one of my eyes, and I have like an irritable bowel syndrome which is linked to it, but I don't know whether it's actually that or not – so, probably no – nothing that's affected my life in any way.

Yes, it's something that comes up a lot and like what runs in families because our family has a lot of irritable bowel syndrome in it and there is some diabetes – it is very common. Sometimes you tend to get these things anyway.

Those are the sort of things you could have anyway. Cataracts – anybody can get them, so you don't know whether it's specific to that or not.

I'm going to ask you a few questions about ageing. Just think about yourself and your sister – you're the older sister. If you think over the last 10 years, going from teenager through your twenties – how do you feel about yourself and your physical appearance?

I'm probably a lot more confident now than what I was in my early twenties. I don't know. I'm a sort of happy-go-lucky person and I just sort of take each day as it comes. I don't really worry too much about my physical appearance. I'm much more confident now. You know, I have just that, like well, that's-just-me attitude and just get on with it.

So, for ageing, I don't know. I'm probably a teen in the head, anyway – so, it doesn't really matter.

If you think about you and your sister – you're both quite stylish. I wondered if you could give me an anecdote about sharing clothes, strange outfits that you've both worn...

Or if you think her dress is completely outrageous?

No, I don't at all. To Laura style really comes naturally. She has had her own style from when she was an early teenager – even from before that. She always knew exactly what she wanted to wear and what she had, and she always yelled at me as being ridiculous, like somebody else, because she just wore it, and pulled it off. It worked for her. I always tend to be a bit more on the conservative side of dressing – so it took me a wee while to find my own skin with it. No, Laura has always had an eye for it. We are quite good at picking clothes for each other. She would say, 'Oh, Claire would like that', and I'm always going, 'Oh, Laura would like that'. 'Oh, Laura would like that handbag'. Probably, secretly, because I kind of like it myself – but she'd have a fit if I bought it.

Do you borrow each other's clothes?

Aye – sometimes. We're quite lucky. I think it is lucky that my feet are smaller than hers and that she's taller than me because it might end up in a bit of a fight if we borrowed each other's clothes. (She says, sitting in her sister's clothes right now.) *[Laughing.]*

When you get a form to fill out your age, what box would you say you're in – if you think of young, middle aged, or older?

If I saw a form with a tick box and it said to clarify your age, I would probably class myself as young.

You said a wee bit about your sister and her style. Can you describe your sister as she is now?

My sister is a 6-foot-2 blonde bombshell. Covered in tattoos, piercings, and – yes, she is just a lovely, big, beautiful girl.

Say a bit about her tattoos.

Laura and her tattoos. I think it started when she was 16 – that was the first tattoo – and Mum allowed her to have a small tattoo on her shoulder. So she went to the tattooist; got her tattoo; came home; phoned me and said, 'Would you tell Mum that I have a lot bigger tattoo on my shoulder than I started with?' I said, 'Why don't you tell her?' She said, 'Because you're in England and she can't shout at you'. And that was the start of her love affair with tattoos. Small ones turned into big ones and she's now got fantastic works of art. She's found a tattooist that she really loves and she's just really defined herself and her style with all her tattoos and her piercings. They're really her; they're just so unique to her – skull and crossbones and pirates and underwater things and all sorts of things. The guy who did them is brilliant.

Whereabouts are they?

She has a massive shoulder piece; she has another one on her shoulder; and then she has a big lower-back piece and she has a side piece – like an orchid garden on one side, and she has a fairy on the other side, and she has a heart on her wrist, and she has a skull and crossbones on her ankle, and she has a ship on her foot. I think that's it, but there's probably more.

Do you have any tattoos?

I have tattoos, yes. I've got three tattoos. I have angel wings on my shoulder, I have a butterfly on my lower back, and I have a peacock on my thigh.

Who was the first to get them?

Laura was the first to get them. Oh, no – I've got a butterfly on my tummy! That was my second but I was about 19 before I got that. It's about the size of a 50p [coin], and I nearly had a heart attack getting that done.

Was it sore?

No, I don't think it was sore. It was more the idea. It only took about 15 minutes and it's not really a very good work of art. I think it was the thought of actually going through with it – that was more the point than getting it done.

Do you have your ears pierced?

Yeah.

You said a wee bit about your style. How would you describe your own physical appearance?

I don't really know. How would I describe my physical appearance? I'm kind of tall and quite thin. That's really about it. I've been trying very hard to become tall and a bit more muscly by going to the gym, but it's sort of had a counter effect. I just got a little bit thinner. That's the way it is. I don't really have any style or anything – nothing in particular – so it's just, whatever suits me and fits me is kind of what I wear.

You've just changed your hair.

Yeah. The mop at the top has gone. I cut all the curls off. I've been threatening to do it for about four years for my long, curly, fuzzy hair, and I just bit the bullet the other day and said, 'Right. That's it. Cut it off'. And my hairdresser went, 'I've been waiting for years for you to say that'.

What was your hair like before?

My hair was the same sort of style. It was about this length *[indicating shoulder length]*; it was thick, dark brown and just is naturally curly so it just kind of... That was it. I don't really have anything else to it.

I want to talk a bit about your mum and your aunt – how you've seen them age, because they're both now in their late fifties or early sixties?

Yeah, Mum is in her fifties, and Elizabeth has just turned 60 last year.

Can you talk about how they have aged in the last 10 years – their faces?

Yeah, I suppose they've both aged fairly well.

I suppose my mum has aged... Well, she has aged very well for all her love of sun and going out... She dresses very stylishly and she has her hair cut well; she always has make-up on when she goes out and sort of tries to do her best to make herself feel good. In the last 10 years – I don't really think there's been a change in her. Maybe the odd style change or hairstyle change, but I think on the ageing process, she's aged well.

My aunt – well, I suppose because of her myotonic dystrophy, she's losing some of the elasticity in her face. Her eyelids are quite heavy – but, apart from that, she doesn't really look much different than she did. She's lost some weight and I suppose that makes her look thinner, but I don't think she looks much different than she did 10 years ago.

To me, I don't know, because they're my aunt and my mum, they kind of always look the same anyway because you just see your aunt or your mum. You don't look at the wrinkles.

Do you ever think about the long term? Your parents have retired now, and you work with elderly people. Do you ever have any thoughts on what the next stage of ageing will bring?

I think for me, I block it out in sort of 10-year blocks – so, this decade I would like to maybe have a family of my own; the next decade bringing them up; the one after that the same; and for when my parents are older, it will be to look after them – so, if they need care at home or need a bit of extra help, that's what I feel my duty is to do for them. So, I just kind of block it out that way.

I don't really think too much about how I will age. I'll just see how things go that way – just see how it goes.

So you said that you'd like to have a family.

Yeah.

Is that something that you thought about in relation to your diagnosis?

Yes, definitely. The diagnosis has been a big factor in that and I think because of the diagnosis is the reason why I don't have a family now, at 31. My husband and I have discussed it. My husband was fostered and he was brought up in a foster-care family, and sometimes foster-carers can adopt which is something I've always thought of, first. As soon as the diagnosis was made, I thought, 'I'll not bother with having a family of my own. I'll stamp this myotonic dystrophy out of my line and that will be the end of it and adopt or go into foster care', but that's something that he's not keen on and he would like to have a family of his own. He would like to have his own child – so we've been in contact with the geneticist and things like that, and that is the process that we're hoping to go down over the next year or so – so, it will be a long drawn-out process. But – yeah – so that's something we've definitely thought about, we've talked about. We never kept it from John. We have always spoken about it – we've had very open discussions about it, so we both know where we stand and that.

I think you take the element of surprise out of it and you have to sort of think about it more... There's so much more planning that has to go into it. We have to go through all the stages and all the appointments and that sort of thing to eradicate myotonic dystrophy from my line. You sort of go, 'Oh, I'll do it next month' because you can kind of choose. It sort of takes the surprise element out of it. Maybe it's a good thing. I don't know.

So you're hopeful to go down that road – but if that wasn't to work out, would you go back to thinking about adoption?

I think I'd definitely go back to considering adoption. I still haven't completely ruled it out. Once we get into the process and see how things are going... I just think there are so many children out there looking for a good home which we can provide for them – but if the treatments didn't work – because it might not even be myotonic dystrophy. It could be any other problems and there's nothing to say...

You're not making a designer baby by taking out myotonic dystrophy. I would definitely look to adopt – yeah. I would love to foster, but my husband doesn't. He's not into fostering at all – so... Going through Social Services – he doesn't want to do it – so...

You said he grew up in foster care?

Yes, he grew up in foster care.

Do you want to say a little bit about that? Is that OK?

Yes, that's fine. John grew up in foster care. He moved in with whom he calls his mum when he was four – him and his brother. He and his brother got the same mum, but different dads, and he was 18 months. They were very lucky to be able to stay with Susan for their entire life. Every six months a care review was done and they remained in their family home. They had a really happy childhood and really enjoyed being with Susan.

I think it was a bit mad at times because she did emergency foster care and there were, maybe, 10 or 11 babies or children in the house at one stage and then, two weeks later, it was back to sort of five of them – but she very much considers them and accepts them as her own children. She stood by them for so long. She's an absolutely fabulous woman – absolutely amazing. She's 76 and she's just like a spring chicken. She's brilliant. She's a credit to us all.

Keeps you young being surrounded by all these children!

I want to go back again to thinking about the book of your wedding and you just look so beautiful in that, so I wanted to talk about what makes you feel good – it could be holidays, it could be getting dressed up, but just talk about what makes you feel really good in yourself.

The best thing that makes me feel really good in myself is a more recent thing. My sister and I both joined a gym at the same time and it's quite hard going, but I love that feeling that you get once you've done a really heavy workout session and you're just lying on the floor and you just feel fantastic, and then you go and have a coffee with your girlfriends and chat and natter and you all look a mess and it doesn't matter because you're just all in the same boat.

Getting dressed up, like going out. I do like it, and I like buying clothes. I like getting dressed – but that wouldn't be the main thing. I like things that make you feel good like going out and having a really lovely walk on the beach with the dogs and the sun poking its head through the clouds. Those are the things that make me sit back and go, 'You know what, this is good'.

So you never used to do the gym? You're all a very fit family...

We've always been fit. We've always walked and talked and played sports. I used to do a lot of swimming when I was younger and I took up rugby. I played rugby for a year for a women's team. I loved it, and then just kind of became converted to the gym thing and it gets me up at half nine in the morning on my days off and I'm in the gym for half nine, and there for a couple of hours and I just absolutely love it – just that feeling fit and fresh and healthy.

Do you worry about a point where you might not be able to do things?

Well, that really has struck my mind. Going back to earlier about wondering whether/how the myotonic dystrophy affected your body – I suppose a lot of the things are very quick and explosive. I would struggle with that. I've always wondered, is that part of myotonic dystrophy, or is that just part of me – you know, some of the gymnastics was hard, or getting off the block for sprints would take me a little bit longer just to start.

I don't know. I think if I just keep myself fit and strong. It doesn't matter what you're doing if you can still swim, and you can still put one foot in front of the other, you may as well just do it.

There are lots of studies that show that that really helps.

I sort of felt that myself. If your muscles are going to waste, and if that is going to happen, it's better off wasting from a higher point than from a sedentary point of view. You know, if you are fit and you are strong... I suppose, more so, it gives you the determination to do things as well and if you are going to weaken, you're probably better starting out as strong as you possibly can be. I don't really know about studies on it – but that's how I work it out in my mind.

Yeah, I think that seems quite a common-sense view to take.

Thinking back to your haircut and everything – you really don't think about ageing. I talk to my sister about ageing a lot and she says she's getting everything done – you know, when you see famous people and they've had surgery or Botox. Do you ever think about that? How do you view that?

I don't judge anybody who's had any cosmetic procedure done. I think it's your body, and you can do what you want with it. I sometimes think it's maybe a self-conscious point of view if people feel the need to stay young, keep themselves looking young. I'm not really that worried about it. I use face cream in the morning and I put eye cream on, and that's about the height of my beauty routine – and I only started about two years ago. I don't really think about my skin getting older and things like that. I don't really worry about it because we've grown up that way.

My mum has beautiful skin and her mother had lovely skin. My dad's mum was amazing. The day she went into hospital (unfortunately she never came back out), she still had her hair set in rollers and had lipstick on. You sort of see beauty on the inside, more than the outside. I don't really think about getting older.

If you could think of somebody that you would describe as beautiful, who would that be? Think of people you would describe as beautiful.

Who would I describe as beautiful? Some people are beautiful... I think Holly Willoughby on TV is beautiful because she's just got such a big beaming smile and has lovely bouncy hair. I think Prince William's wife – she's beautiful because she just glows.

Do you think your sister is beautiful?

Oh, yes. She's lovely. I think Laura is very pretty, because she just has that lovely porcelain skin and because she wears her make-up really well, and even without make-up on, she just has that lovely... I suppose our family trait is a round face. She's got a round face and lovely skin and dark eyebrows and dark eyes. And if she didn't have her hair dyed blonde, she's got lovely dark hair too.

I was going to ask you one more: going back to your mum. Your mum says she's had her hips done. To just maybe talk about that process of what that was like when she was having that done – about your mum being in hospital because she was in Glasgow for one of the times?

Mum had both of her hips replaced. The first one she had resurfaced where she got part of the bone at the top, I think, they put like a sheath over the top of it because she's got osteoarthritis, and it worked really well and she recovered well from it. I think it was harder on her than it was on us because we expected her to maybe not be able to do things the same whereas Mum has more like a got-to-get-up-and-go attitude. She just wanted to go straight in, and she's just had her second hip replaced there last year and I think it was a bit harder on her. She didn't adjust as quickly as she did the first time. It took her a little bit longer to get back to normal. She needed a hand with getting up, and a hand with getting dressed and things, which I do everyday anyway, and it took no toll on me whatsoever, but it took her a wee while to recover from that one.

There was a classic moment when, including Laura and myself, when Mum was recovering from the first one, and we all decided to go out shopping, so we put Mum in a wheelchair. She got quite cross because we kept walking off and picking stuff up, 'What do you think of this, Mum?', and she said, 'I can't see, I'm in a wheelchair!' That was quite a fun day out.

My sister sometimes uses a wheelchair when she's going through airports.

When she's going out, yeah. We started using a wheelchair on trips for my aunt. We all used to go out with a choir; we'd be gone for a couple of days and Elizabeth couldn't handle a lot of the walking, and so we would put her into the wheelchair and then she would be our bag carrier – so everybody just put all their bags onto the side of the wheelchair and Elizabeth carried them all until one day we went for coffee and Elizabeth sat in the chair and the wheelchair fell over because of all the shopping on it. But I think it helps her. It's much easier on her. When she goes away on holidays with Steven, I know they use the wheelchair to transfer her onto the plane but fortunately Elizabeth has grit and determination that that is not what's going to become of her. She's still on her feet – so it just makes those long journeys a lot easier for her.

She's very determined...

Yeah.

Talk about your Aunt Elizabeth's driving.

My Aunt Elizabeth's driving is atrocious. I don't even think it has anything to do with myotonic dystrophy. I think it was bad to start with. It's luck, not judgement, that gets you somewhere. That's the polite way to put it.

You're at an exciting time in your life – thinking about, if you had a magic wand and could change one aspect of myotonic dystrophy in the way it is looked at or researched, what would that be?

If I could wave a magic wand and for everything to be OK with myotonic dystrophy, I would probably, first of all, I would educate, because everybody could say, 'Let's get rid of the disease' and that's like gone but, as we know, it doesn't really matter if you stop it, if you find out you have it and you stop it at birth, or you decide not to have children and don't pass it on – the genes do mutate anyway, so you never know if it's going to come back but I suppose education, more than anything else. A lot of doctors don't know what it is. A lot of people don't have a clue.

Obviously I'd like it to go. I'd like them to find a cure. Well, not a cure, but to be able to screen it out, but more so to make the lives of people who have myotonic dystrophy a lot easier. Everybody goes, 'Oh, myotonic dystrophy, I know what that is' – but they don't. They think somebody's drunk because they speak with a slur. They ask you to sign forms and it's not on it. I don't know – educate people, and then, hopefully, that would make it a lot easier.

Laura, 28

Northern Ireland, 7 February 2014

Jackie: *If you could tell me what your name is, and how old you are?*

Laura: My name is Laura, and I am 28.

And a little bit about where you were born and grew up.

I was born in Belfast. I grew up in Belfast. I lived there until I was 18 and then I moved to Glasgow for university and stayed in Scotland for nine years and came back to Northern Ireland almost two years ago.

I would like you to tell me about your childhood and what your schooling was like – where you went, what type of school it was – a bit about that.

I went to a very generic primary school. It was nice. It was good fun. Secondary school, I went to an integrated school; in Northern Ireland, there was usually separate schools – so I went to an integrated school which was a mixture of Catholics and Protestants and it was also a mixture of a grammar stream and a comprehensive – so, it was a good school. You got a really middle of the way education; you got to learn about lots of different faiths and lots of different communities and stuff. It wasn't one-sided at all. I stayed there until I finished my GCSEs [exams]. Yeah, it was OK.

I didn't really like school very much. I didn't like the authority of being told that I couldn't wear purple hair and have piercings and things – but I did OK at school. I finished when I was 16.

Can you say a little bit about what you studied at university?

Yes, I studied history and sociology for my undergrad at Glasgow Caledonian, and it was good. I have a postgraduate diploma in local history as well which I did with the Open University after I'd finished uni. I love history, so I studied history all the way through GCSE, A levels [exams] and then degree as well and then postgraduate – so, hopefully, history will lead the way for the rest of my life.

What particular periods in history?

I am really interested in Victorian, Edwardian, and industrial history – things from the 1750s onwards, changing Britain, and the rise of industry, and the rise of big cities and all the horrible things that come with it, like poverty and prostitution and death and all the really grisly side of history, and particularly social history.

I love learning about different people and how they lived and the different classes and differences between upper classes and working classes and poverty laws and things like that. It's a bit dirty, but it's quite good. It's very interesting learning about it.

And what do you do now?

I work in a museum in Belfast called The Ulster Folk and Transport Museum. It's based on 1900 to 1914 – so, it's a costumed role that I have, so I'm dressed as a Victorian maid. I don't get to be a lady very often. I will take school groups and teach them how to bake soda bread on a griddle on an open fire, or I'll teach them how to make rag rugs out of scraps of fabric. We do crafts with school groups that come to visit and make flags from the *Titanic* era and things like that. It's really good fun and you're really getting them to engage with history and they come away not realizing that they've actually learned quite a lot. We try to make it fun for them. It's really a cool job because people get to learn about history without actually realizing that they're learning it. It's not a classroom environment so it's a really good job.

I'm going to ask you some questions about your family situation. If you can talk about whether you have any siblings and what your position is in the family?

I have one sister. There are just two girls in our family and I'm the youngest – so my big sister Claire is 32 (I think). Yes, it's just the two of us and Mum and Dad.

I want to ask a wee bit about what you know about myotonic dystrophy, and just explain anything that you know about it.

Myotonic dystrophy, I know, is a muscle disorder. It's a muscular dystrophy involving lots of muscle wasting and gets quite bad as you get older. I know that there's lots of different forms of variables and levels.

Myotonic dystrophy is a muscular disorder. It's a genetic disorder that runs in families. It can be quite debilitating from the word go or some people have it where you wouldn't even know that they have it, and it varies from person to person and family to family. It can be quite bad in some respects, and in others, if you keep yourself fit and healthy and eat well, then it can be OK.

Could you say a wee bit about the process where you found out about it in your family and the decision that you made with your sister to go and find out more?

The process for us about myotonic dystrophy – we knew about it from when we were children. Mum never pressured us into getting tested. They didn't know about it until she had already had it. Myself and Claire, we made the decision when I was about 18, I think, to go and be tested – so Claire and I went off to the hospital.

It's not always a thing that you think about…

No, we tend to, because it's in our family. It does not really affect us on a personal level so much because Mum and Claire are so good. We just kind of ignore it. It's in the background, really.

My sister and I decided, when I was 18 and she was 21, that we would go and get tested. To be honest, people thought I had it, and Claire didn't because I always was a very sleepy child and would have had lots of colds and flu and not as much energy as my sister, so it was quite a surprise to find out that I didn't have it, and Claire did.

I'm also a very emotional person, as you might find out, and Claire isn't. You would think that with myotonic dystrophy I would be a bit stronger because I don't have it and Claire wouldn't be, but it's the other way round. Sorry [*upset*].

That's actually, exactly the way round. It's the same with me and my sister, she's never ever got upset about it – not once.

No – not once.

Totally calm, and it's me that goes flapping about. It's very common.

It's so good to get people's perspective on it. It's really important that doctors understand what that process is like for people. They just see you for 20 minutes, and they do a test. I know you have great doctors here, but it's really important that people see what a seismic thing this tiny appointment can lead to, which they don't always take on.

The way we found out, I went to the hospital with Claire and she had to go and have her heart checked on there and then, but Claire has a very mild form, I guess. She wasn't too upset. I was quite upset, but she was ok. Since then we just got on with it. We try and stay fit and healthy as a family because we know that that will help to keep her muscles nice and strong and that's really it. We just get on with it.

You are a very fit family.

Yes.

Can you talk a bit about how important fitness is to your family?

I understand how important it is to stay strong when you have myotonic dystrophy, to stop muscle wasting, or to prevent it as much as you can. My mum has been a great role model and the fact that she's always been exercising and active and even now, nearly 60, she's still a fitness instructor and is really inspiring, so me and Claire go to the gym together. When I came home from Scotland I dragged her to our local cross-fit gym. Before then, she was running quite a lot but she didn't do a lot of weights, whereas I have already done quite a lot of strengthening and conditioning work so I dragged her to the gym and we started lifting weights together and she's got really, really strong, and so have I. It's great. We're both fit and healthy and Claire is probably the fittest and the strongest that she's ever been. She knows that if she keeps that up then she has more chance of keeping myotonic dystrophy at bay.

Because it's been in your family, how have you viewed myotonic dystrophy in terms of your mum's family and this sort of legacy idea? Is it something you thought about, or not at all, until you went to get tested? Do you think it coloured your childhood?

No, I don't think myotonic dystrophy coloured my childhood at all – but, again, my mum and my sister have quite a mild form and even though my auntie has it quite bad, she's still able; she still goes on holidays; you can still speak to her on the phone; she's not in a wheelchair; she's not very debilitated by it, although she is getting worse as she gets older. So, I don't think it stopped us from doing anything as a family. In fact, it probably brought us together and kept us nice and close because we knew we had this bond, and we had this thing that is important to stay close as a family.

I'm the youngest sister. My sister Claire is a few years older than me. We had a great relationship when we were small, little kids. We used to play together and go on holidays together and I have no bad memories of fighting or anything when we were small children, but we had a bit of an edgy teenage years where we didn't get on very well.

She moved away when I was 15 and Claire moved to Portsmouth, and then I moved to Glasgow – so we didn't see each other for quite a substantial amount of time. It was always a bit fraught whenever we were home. We used to argue a lot and I would just take myself away for a few days if she came home – but I also understood that she wanted to spend time with Mum and Dad on her own, so I would take myself off so that she can have time with them.

But as we got older, we have quite a good relationship now. If we see too much of one another we might get a bit annoyed, but we would go to the gym together which has really brought us closer and we would go and drink wine on a Saturday night and go shopping – so, I definitely think that as we get older, our relationship is getting a lot better and a lot stronger.

I want you to think about yourself. Do you think you look like your sister? Is there any family resemblance?

I think when we were younger, we didn't really look like each other. My sister Claire was blonde and I had a full head of black hair when I was born. I grew to be very tall, so I outgrew Claire when I was a teenager. I still don't think we have much of a resemblance, although people tell us now that they can tell we're sisters – but my sister Claire looks very like my mum and I tend to look a bit more like my dad. You would know we were sisters if we were out with my mum because we're three tall girls and you can see a family resemblance, but if we were just out, me and her on our own, you wouldn't really know. In fact, when we joined the gym together, it was about six months before people actually caught on that me and Claire were sisters. They thought that we were just good friends who were going to the gym together – so, I don't think we look like each other in terms of facial resemblance and because I'm taller than her as well – we don't have the same height or the same hair or anything, but when you see Mum and Dad with us, you can tell that we're sisters.

Do you notice any differences from how you look, how age has affected you, and also about your style compared to each other as you both get older?

Probably the main difference I can see between me and Claire being teenagers and now is our hair. Claire used to have short, straight hair, and now she has very thick dark, curly hair. My hair changes all the time, so I can't really use that as a guideline.

I think we both quite stand out. Claire wouldn't be as brave as me in terms of clothes and tattoos and hair and piercings and things like that, but she can be very stylish as well whenever she goes out and she always looks great whenever we go out for a night out – things like that.

Could you tell me a bit about Claire getting married?

When Claire got married four years ago, she had a bit of a war theme wedding. I don't think it was intentional. I think it just happened that way because people were lending her marquees and cars and things and she had a fabulous fifties-style, and full circle, great wedding dress; had her hair all done in rollers and pins and it was lovely so I kind of followed it with a fifties-style dress as well and I had bright red hair and my tattoos out, and a big flower in my hair and things – so, in terms of traditional weddings and what they look like, it definitely wasn't one of those. It was a great day.

Did you ever share clothes when you were teenagers?

I don't think me and my sister Claire ever shared clothes. I was a bit of a punk and she was studying to be a beauty therapist, so Claire was a bit of a full-face-of-make-up and always had her eyebrows perfectly waxed and nice clothes whereas I was a bit of a grungy punk with black hair and black nails and tie-dye clothes – so, no, we never really shared clothes! Although now we would share shoes quite a lot, if we can. We seem to have a bit more in common in terms of fashion now as we are a bit older.

Your style is very distinctive. Would you say that it has evolved over the years?

I would say my style has evolved. I was always a bit different. I don't care about following trends or crowds. I'm nearly 6 foot 2, so I always stood out – so, to me, I guess, it was just normal. I didn't want to fit in with everybody else because I already didn't fit in, so I just thought, 'Well, sure, I like this rock and roll style; I like wearing black and going to punk gigs' – so that's just what I did.

As I got older, I'm still not the same as everybody else; I still have my own sense of style. I would put on clothes other people wouldn't even think of wearing. I'm covered in tattoos, piercings, I like having pink hair sometimes, even though I'm nearly 30. I don't think I'll ever change.

My mum hoped I would grow out of my grunge style when I was a teenager, but it never really happened. But now, I think, she's quite proud that I can be different and be myself.

Could you tell us about your tattoos; how many you have?

I have about nine tattoos altogether. I love colour, so I have lots of colourful tattoos. I have flowers, lilies and things on my side; I have an underwater theme on my arm with an anchor, an octopus, and two sharks and a treasure chest and things. I love being in the sea, so I love everything nautical. I have a pirate ship on my foot; I have a Jolly Roger skull and crossbones on my ankle; I have a few others on my back and I have a little flower on my wrist – but mainly colour. Colours are just really nice and pretty and I don't like to have tattoos that other people have so I get them all hand designed with an idea and don't just go in and pick them off the wall. Hopefully nobody will ever have what I have.

Would you say, at the stage that you are at now, that you're happy with your body?

I think I am happy with my body. I would say I would like to change a little bit and tone up a little bit more around my tummy, but I lost quite a bit of weight, so I am quite toned. I have quite strong muscly legs and muscly shoulders and things, and I'm quite happy with that. Being 6 foot 2 I'm never going to be stick-thin and I quite like having curves – so I would say I'm quite happy with it, yeah.

What makes you feel really fantastic? What is your best look, if you're getting ready to go out, or what makes you feel really, really good?

I love getting dressed up in vintage fifties style – either wiggle dresses or a nice big full circle skirt or a nice pencil skirt and a pair of heels, and I would curl my hair and put little pin curls in it and bright red lipstick and fake eyelashes and things. I don't do that very often, but I do enjoy getting to go full-on when I go out and I think I can carry the style off quite well, especially with my tattoos. It looks quite pin-uppy. That would be my favourite way to get dressed up.

You talked to me about going to the gym with Claire and lifting weights. What do you think you get out of your sessions with Claire and how do you feel at the end of it? Can you talk about what the word 'strong' means to you in relation to physical strength?

I absolutely love weightlifting. I think everybody should lift weights. I don't think a skinny look is a good look. I think fit and healthy is a far better look than looking stick-thin and looking like you've starved yourself and the gym we go to is definitely about being strong. It is not about being skinny. We're encouraged to lift weights in a proper way and you can see your strength gains go up every week, almost. When you start off you record what weight you can lift in certain things, and a year and a half later you can see such a change. It is quite exhausting, but it's a nice feeling, as Claire said. You get to the end of a session and you know you've put maximum effort in and you can see and feel the benefits of it – so that's why we keep going back, and that's why we both love it so much.

Have you ever had to lift Claire or your aunts – even as a joke?

No. I haven't had to physically lift people but I know that Claire does in her job, and I know that my mum would help out with Aunt Elizabeth if they're away for weekends and things. She would help. I have helped Elizabeth get out of chairs and things like that and helped her into the car and stuff, but I've never actually had to physically lift her from the ground up – but I think I probably could if I had to – hopefully.

That is quite interesting in Claire's job.

She'll be fine. They're going to set up a nursing home for Mum and Dad and Claire's going to look after them!

The next thing I want to talk about, have you noticed your parents ageing? Is it something that you've become aware of?

To be honest, with my parents getting older... They're still so active, there's not really a huge difference. I know that my mum doesn't do as much high-intensity exercise anymore, but she still would do Pilates and go walking and my dad is out walking nearly every day. We've always been a very active family and I don't think that will change.

I have noticed with my mum, her knees get sore and her hips get sore. She's had a couple of hip replacements, and that was quite hard for her to sit and do nothing for six weeks and then have to recover, because she is so active and she is usually out and about and not really in the house very much and always at work or always teaching and things. That was quite hard for her. I think she probably finds that difficult as she gets older, to not be able to do as much – but, at nearly 60, she still does more exercise than your average 21-year-old, so I think she'll always be quite active.

Can you talk a wee bit about what you think about your future, if you have children, or if you'd like to...?

I would like to get married and have kids. That's certainly something that I hope to fulfil in my life. At the same time, I'm quite happy on my own despite my mother's intentions of getting me married off by the time I'm 30 which I doubt will happen (by that time, anyway).

But I would like to meet somebody and get a house and have some kids, but I have decided that if I don't meet anybody, I am going to have children anyway and I have a lovely gay friend who I have spoken to about it, and we have decided that in five years' time, if I still haven't met anybody, and I still want a child, then I'll have one.

What do you think about getting old and if you have any fears?

Do I have any fears of getting old? No, I have no regrets in my life and I hope that it stays that way. I like having fun and I hope that, for the rest of my life, I can continue to do that and get to do old age with a walking stick and being able to tell my grandkids about all the rock-and-roll stories that I've had in my life. That will do.

Would you consider any anti-ageing treatments like Botox?

I don't think I will ever get Botox or anti-ageing injections or facelifts or anything like that. I think exercise, eat healthy, drink plenty of water, and you don't really need them. I don't like the fake plastic look that a lot of girls have and I don't tend to be paranoid about getting older, having wrinkles or anything like that – plus, you don't know what you're putting in your body and I don't think I would like to put a substance that I'm not a 100 percent certain of in my system – so I definitely don't think I'll ever get Botox although, I'm only 28. Ask me again when I'm 38. It might be a totally different story!

I know my sister Claire has never mentioned to me about having Botox or anything and my mum has never mentioned facial cosmetic surgery anyway – so, no, I don't think I would have anything like that. Just try and remember to take your make-up off at night and drink plenty of water, and hopefully that will work for me.

How do you feel about shoes? What I have asked everybody else is whether or not they enjoy wearing high heels.

I love shoes – all sorts of shoes! Flat ones, high ones, sparkly ones – I love them. I do wear high heels sometimes when I go out, but not very often because I'm incredibly tall anyway. If I put a pair of high heels on, I walk into a bar, and I'm 6 foot 5 or 6 foot 6, I'm taller than every single person in the bar – so it depends how I'm feeling at the time.

If I'm going out with new people, I tend to not wear them because I don't like getting comments about being so tall, so I'll tend to wear flat shoes and then once I get to know people, I'll start to wear them again.

I never owned a pair of high heels in my entire life until I was 22 or 23 and then my friend asked me to do some vintage modelling for her because I was so tall, she thought I could carry it off and she made me wear high heels and from that moment on I have absolutely loved them – so now I have a nice collection and they sit on my floor, and I admire them quite a lot. Sometimes I put them on and walk around the house, and take them off again.

I do like to wear high heels. I think they definitely make you look a lot taller and slimmer and really add to your figure – especially if I'm going out in a nice dress to a wedding or anything, then I would wear them just to finish off the outfit.

The last bit, if you had a blank cheque for any kind of cure or treatment or process or anything that you could do within the world of myotonic dystrophy in its broadest sense – or something that you could just change immediately, or an appeal that you could make, what would it be?

I think my appeal for the world of myotonic dystrophy is to try and encourage people who have it, to have children through IVF so that it can be stopped and it doesn't have to filter down through generations. I think that would be the biggest thing for them because then you're not having to make a decision if you have a child and it's disabled, you don't have to make that decision. You can deal with that before you have a child so that you can have a child who's born without it and also continue through your family line without passing myotonic dystrophy through.

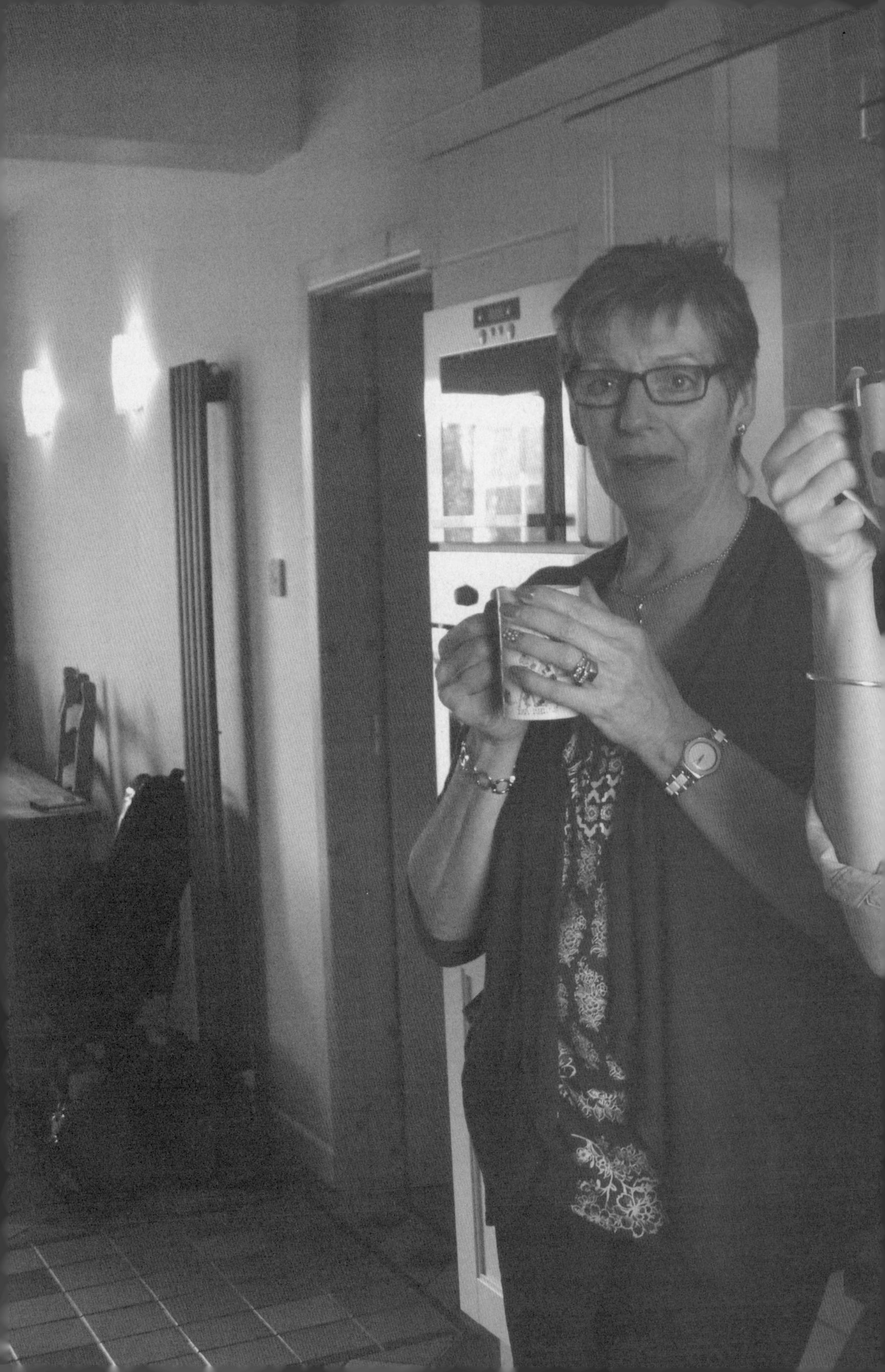

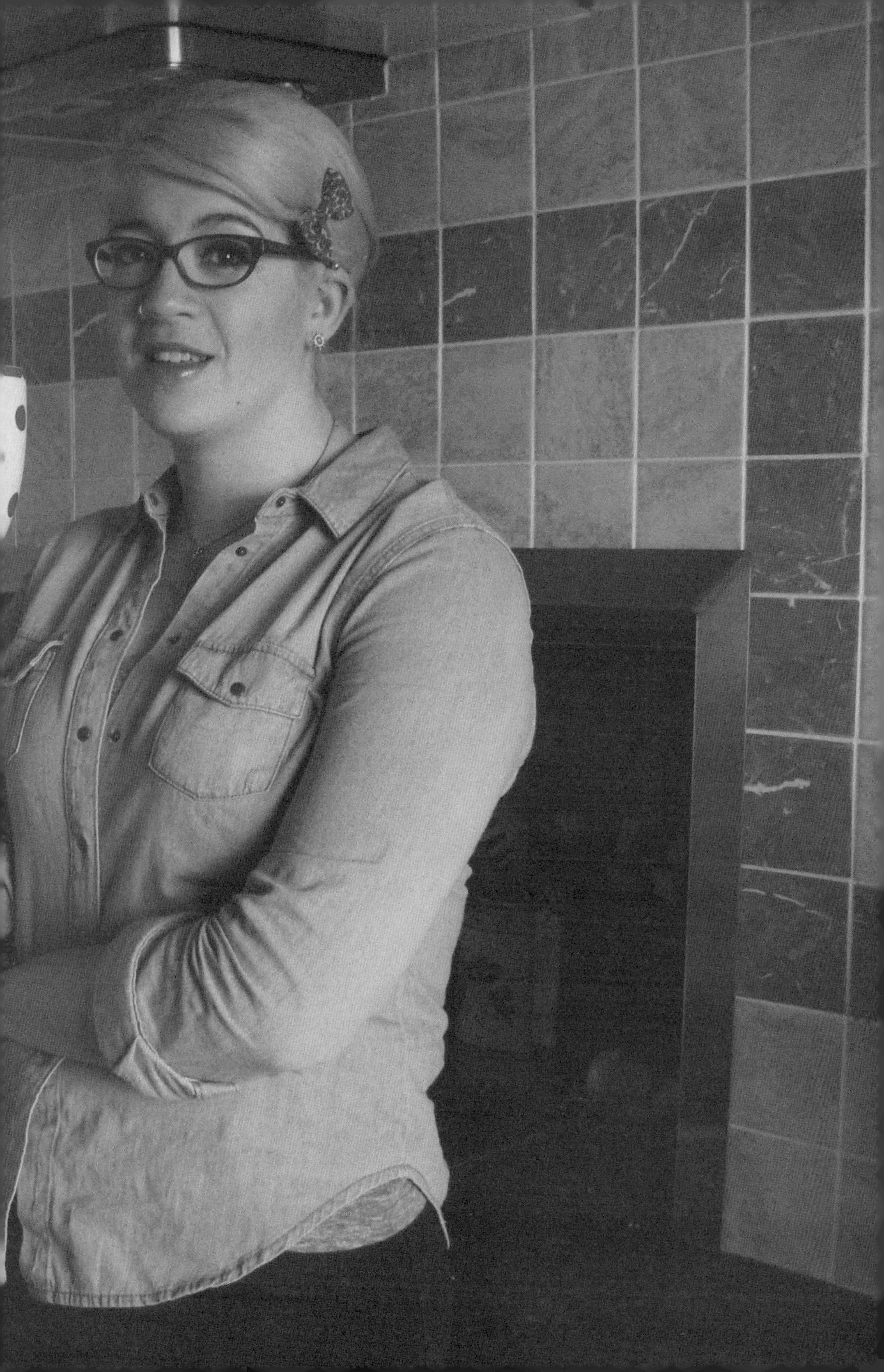

Further Information

Myotonic Dystrophy

Myotonic dystrophy is a form of muscular dystrophy, the umbrella term that refers to a whole group of disorders that cause progressive weakness in muscles. It is the most common form of muscular dystrophy in adults, a complex genetic condition that can cause symptoms, from mild to severe, across three generations of a family due to its inheritance pattern. This makes it a fascinating scientific subject, but causes a significant burden to individuals affected, and their families.

It is a dominant inherited disorder that can come from either a mother or a father, though the severest form – congenital myotonic dystrophy – is usually only inherited from an affected mother. It occurs because the gene has mutated in such a way that it functions differently, and any affected individual has a 50 percent chance of passing the condition on to their children.

It is often viewed as a muscle disease, therefore those affected could have skeletal muscle problems, heart function abnormalities, breathing difficulties, or issues with speech and swallowing. Also common are cognitive impairment and excessive daytime sleepiness. The variability means that some individuals could experience all of these symptoms, some only a few. Many symptoms, such as fatigue or irritable bowel syndrome, are also common in the general population, and this can lead to under-diagnosis as symptoms are not always attributed to the condition by general practitioners.[1]

As myotonic dystrophy is passed on, symptoms generally appear earlier, and are usually much more severe; for example a grandfather may have had very mild symptoms such as cataracts or muscle stiffness that only appeared late in life, but his daughter or son may develop more significant issues with walking, swallowing, speech and breathing in adulthood. The third generation (with congenital myotonic dystrophy in particular) may have severe problems with low muscle tone, ability to walk, speak and eat and will usually have significant learning difficulties from birth. Often, families only become aware of the condition at the birth of this third generation, as many infants are born prematurely, with breathing difficulties. This inheritance pattern, known as anticipation, can then lead to a three-generation diagnosis being made in a family within a short space of time.

And so it is a condition that causes a number of notable difficulties that can include physical and psychological disability and a severe decline in mobility, speech and appearance. As muscle weakness is a significant symptom, it can also cause some distinctive appearance-related issues – drooping eyelids and a 'blank' facial expression due to low muscle tone in the face, for example. Premature balding in men, and sometimes women, is also very common. Symptoms can lead to social isolation and a fall in economic circumstances, and a major frustration for families is the lack of awareness and understanding of the condition and the loss that it carries physically, socially and emotionally.[2]

The condition is progressive with no known cure, though following a decade of unprecedented scientific research a small number of clinical trials are underway.[3]

Further information:

Peter Harper, *Myotonic Dystrophy: The Facts*. Oxford: Oxford University Press, 2002, 2009.

Support groups and charities, UK:

Myotonic Dystrophy Support Group
19–21 Main Road
Gedling
Nottingham NG4 3HQ
Phone: 0115 987 5869
Helpline: 0115 987 0080
Email: contact@mdsguk.org
Website: www.myotonicdystrophysupportgroup.org

Scottish Muscle Network
Meridian Court
5 Cadogan Street
Glasgow G2 6QE
Phone: 0141 300 1424
E-mail: laura.craig1@nhs.net
Website: www.smn.scot.nhs.uk

Muscular Dystrophy UK
61A Great Suffolk Street
London SE1 0BU
Phone: 020 7803 4800
Email: **info@musculardystrophyuk.org**
Website: www.musculardystrophyuk.org

International:

Myotonic Dystrophy Foundation
1004A O'Reilly Avenue
San Francisco, CA 94129, USA
Phone: toll free (US only): 86-MYOTONIC (866-968-6642)
Direct: 415-800-7777
Email: info@myotonic.org
Website: www.myotonic.org

1. Myotonic dystrophy affects 1 in 8,000 of the general population in the UK, but this is thought to be very underdiagnosed. It is the most common form of muscular dystrophy in adults.

2. Laberge, L., Mathieu, J., Auclair, J., Gagnon, É., Noreau, L., and Gagnon, C. (2013). 'Clinical, psychosocial, and central correlates of quality of life in myotonic dystrophy type 1 patients.' *European Neurology*, 70(5–6), 308–315. doi:10.1159/000353991

3. A good summary of current clinical studies can be found on the Myotonic Dystrophy Foundation website: http://www.myotonic.org/study-trial-resource-center

Deep in the Heart of Your Brain

20 May–13 November 2016
Gallery 4, Gallery of Modern Art (GoMA), Glasgow, Scotland

Deep in the Heart of Your Brain was a major solo exhibition by Glasgow-based artist Jacqueline Donachie, which included an ambitious body of new work exploring disability, care and loss. This commission for Glasgow Museums, supported by a Wellcome Trust Arts Award, incorporated sculpture and drawings Donachie had made in the last five years with new works developed out of a period of research with a group of women affected by myotonic dystrophy in collaboration with the UK Patient Registry at The John Walton Muscular Dystrophy Research Centre in Newcastle.

In November 2016, a symposium was held at Platform, Glasgow, to develop the themes of Donachie's exhibition. It was co-produced by Katie Bruce, Producer Curator at GoMA, and Jacqueline Donachie, and chaired by Alison Stirling, Projects Director, Artlink, Edinburgh. The aims were to bring artists, institutions, academics and interested individuals together to discuss care, bravery, lived experience, autoethnography and expert cultures in relation to the ethics and practice of knowledge exchange/public engagement in the art/medical research field.

The contributors were:
Jacqueline Donachie, artist: *Illuminating Loss*
Karen Guthrie, artist and filmmaker: *The Closer We Get*
Sarah Cunningham-Burley, Professor of Medical and Family Sociology, Assistant Principal, Research-led Learning and Dean of Molecular, Genetic and Population Health Sciences in the College of Medicine and Veterinary Medicine, University of Edinburgh: *Making the Private Public: Lived experience of health and illness.*
Jason E. Bowman, artist with a curatorial practice, writer, researcher and educator. MFA: Fine Art Programme Leader at the Valand Academy, University of Gothenburg: *Curating as Care Making.*

The exhibition, public programme and the symposium were commissioned by Glasgow Museums with support from a Wellcome Trust Arts Award. Jacqueline Donachie also received support to work with GoMA from the National Lottery through the Creative Scotland Open Project Fund.

The research for, and filming of, *Hazel* were supported by Northumbria University and the Arts and Humanities Research Council (AHRC).

List of works from *Deep in the Heart of Your Brain*
Jacqueline Donachie
20 May–13 November 2016

Winter Trees IV, 2016
Powder-coated aluminium, silk cord
Structure in 2 parts: 4.12m x 60mm; net: 4m x 3m (dimensions are variable)
S.484

Winter Trees V, 2016
Powder-coated aluminium, rope, wool, string
3.5m

Hazel, 3-screen installation, 2016
HD film, running time 9 minutes 48 seconds

Deep in the Heart of Your Brain is a Lever, 2014
Powder-coated aluminium
Overall: 4.43m x 200mm x 4.5mm
Bought by Glasgow Museums from Patricia Fleming Projects, with assistance from the National Fund for Acquisitions, 2017
S.483

Glimmer I, 2013
Pencil on paper
Framed: 1m x 1.9m

Glimmer II, 2016
Pencil on paper
Framed: 1m x 1.9m

Glimmer III, 2016
Pencil on paper
Framed: 1m x 1.9m

Glimmer IV, 2016
Pencil on paper
Framed: 1m x 1.9m

Glimmer V, 2016
Pencil on paper
Framed: 1m x 1.9m

Nice Style, 2016
Powder-coated aluminium
1.8m x 40mm (dimensions are variable)
Bought by Glasgow Museums from Patricia Fleming Projects, with assistance from the National Fund for Acquisitions, 2017
S.482

Studio 1995, 2016
Framed digital print on Hahnemühle Photo Rag 308gsm, edition of 10 + 1 AP
Unframed: 1.1m x 1.8m
Bought by Glasgow Museums from Patricia Fleming Projects, with assistance from the National Fund for Acquisitions, 2017
PR.2017.3

Pose Work for Sisters, 2016
Video, edition of 3 + 1 AP
HD film, no sound
10 minutes 1 second, looped
Install specifications are variable
Bought by Glasgow Museums from Patricia Fleming Projects, with assistance from the National Fund for Acquisitions, 2017
AV.2017.2

Headphone Music, Boats and Trains, 2016
Aluminium, steel frame, plywood, paint, silk cord
4.3m x 5.2m x 2m approximately
Commissioned by Glasgow Museums, with funding from a Wellcome Trust Arts Award

Leg defence, 1500s
Steel
Glasgow Museums, RL Scott Collection
E.1939.65.u.[11]